Contents

INTRODUCTION

FRANKLIN'S SPOKES

In the Spring of 1976 Astronomer Fred Franklin at the Harvard College Observatory in Cambridge Massachusetts saw something everyone would say was impossible. In the nine-inch refractor telescope he was observing Saturn with, he saw that Saturn's B ring had radiating spokes running outward in them, more important they didn't change!

We know that the inner rings travel faster than the outer rings so they should shear apart gradually stretching out and then disappearing, but they didn't. One would think that a discovered oddity, totally at odds with physical laws, would get some notice, but every journal turned down Franklin's paper holding that this was impossible. His fellow astronomers derided his claims, and a noted American Astronomer was cast out by the scientific community.

Even more astounding was the fact that not one fellow astronomer turned their telescope to see if they were actually there. They criticized but they never checked. Every astronomer and institution on the planet Earth chose to be ignorant of the truth. They could have seen them but they chose to not look.

In 1980, however, NASA did look, but not by design. Voyager 2 had finally reached Saturn and began taking photos, and there, in those pictures were Franklin's spokes. The impossible was real, everyone in Astronomy was wrong. But Why?

Why didn't someone look at Saturn?

How is it possible that on a planet full of Astronomers with working telescopes, not one of them took a look at Saturn.

ASD IS MORE COMMON THAN YOU THINK

Many people afflicted with ASD (Autism Spectrum Disorder), share a common trait, they hold fast to repetitive behaviors. Like Dustin Hoffman's character "Raymond" in the movie RAIN MAN, they must repeat certain behaviors. This is probably a coping mechanism that gives the Autistic person some comfort and stability in a world they cannot adequately understand. We know it is a coping mechanism in normal people who do it, and everyone does it. Yes, we the normal have similar behaviors to the Autistic, in science we call it Dogma!

Science is filled with Dogmas and the people who hold to them, and hold tight to them they must. Even when they do look, they won't believe it. They come up with incredible explanations to refuse the truth.

"The Earth is the Center of the Universe."

"The Planets travel on crystal spheres that form perfect circles."
"God doesn't play dice with the Universe."

"Saturn can't have stable spokes."

"Carbon Dioxide is Warming the Planet."

Max Planck said:

"Old Scientists hold to old ideas. Science advances one funeral at a time."

There was a time, as the 21st century was starting, when if you put my name into a Google search, I came up first. No Sex researcher, no economist, my papers came up first:

INERTIA AS A CHANGE IN ENERGETIC STATES, AND ITS EFFECT ON THE TWIN PARADOX,

and;

STRETCHED AND NEAR-ZERO SPACETIME, A NEW MODEL OF GRAVITATION AND BLACKHOLE DYNAMICS.

The most interesting thing to read in the world was not sex, but Physics! I beat out the sex research of Masters and Johnson! The world was not right.

People were more interested in Physics than sex?

Who said that?

Yet, there they were, my Physics papers were #1 and #2 in the internet age.

I would never have thought Physics could have beaten out sex, but it did.

Everyone was reading my papers and why shouldn't they? I explained the cause of Inertia, and that it caused the Time Dilation that allowed the rocketship twin to age much slower than his twin on Earth.

I explained why Black Holes went BANG and created the known universe!

I predicted and explained why the Universe was expanding at an accelerating rate of speed and not at Hubble's Constant. This was big news.

But the world of science was changing and in a very bad way.

On Science blogs around the world, scientists were complaining that funding was drying up. You couldn't get funding unless your project proved man-made carbon-dioxide warming theory called "Anthropogenic Greenhouse Warming" (AGW), and woe to any scientist who's papers said anything that remotely may be seen as challenging AGW. Scientists were being blacklisted, shut out of publication, some lost their funding, their posts, and their jobs. Science was under attack and science was losing.

When I was a nobody, I never really took notice of what was happening. Now that I was someone who was listened to... I felt I had to say something. That was a BIG mistake!

It wasn't my fight really, but I remembered a Jewish man who survived Hitler's Nazi cleansing. He said as a boy when the Nazi came and took away the other groups of people first: the Blacks, Gypsies, Intellectuals, etc. They did nothing because they weren't part of any of those groups. Then they came after the Jews, and he remembers thinking as the Nazi were taking him and his family away:

"Why doesn't some do something to stop this?"

No, the thought Nazi weren't coming after me then, but they would eventually have gotten around to it. Evil grows when left alone!

So, I started to write papers opposing Carbon dioxide global warming. Papers that pointed to the impossibility of man-made greenhouse gas warming of the Earth, and the impossibility of a runaway greenhouse effect. All firmly established by natural laws prohibiting such things. A storm quickly arose up against me. In less than two weeks, my papers were quickly blacklisted and taken out of Peer Review, and I was barred from publication. Years later, the Nobel Prize for my discovery went to the astronomers that confirmed my work, not to me. My contribution to what was called

"The Greatest Discovery of the first half of the 21st Century" went unspoken.

The Autistic by choice crowd was not amused with the Truth, and this is the Truth. If you do not like what you are about to read, you are in good company, but I tell you this truthfully, enjoy it, enjoy your Golden Time because Science always wins out in the end. Nazi win battles, but not the wars.

Science will win this war, it always has, it always does, it always will.

I Was A Strang Kid

I didn't think I was, but adults did.

When I went into the hospital to have my wisdom teeth taken out, the Charge Nurse wrote into her notes that I was a "strange kid".

In my early 90's I was talking on the phone with an old girlfriend whom I dated in my early 20's for 3 years. She said:

"I want to apologize for thinking you were strange."

She never told me that and I was wondering why and asked her. She said:

"When we were dating, you'd see these connections that I just couldn't get, and I thought "Nobody is that smart". But there are these computer programmers at the bank, these guys are changing how we do banking. These guys are really smart, and I realized... people are that smart. You're that smart."

So, if you ever had a college-educated girlfriend from your 20's who said that to you...then you were probably a strange kid too.

Ever since I was a kid, I saw patterns that other people didn't see. I made connections between data sets that others thought were nuts...and so did I. For years I thought I was wrong.

"How can I be right, I'm a nobody, I'm not a genius?"

I'd say to myself. Even when my own experiments proved I was right, I thought I was wrong.

Then, on January 25, 1983, the Infrared Astronomical Satellite, the first space telescope to map the entire night sky at infrared wavelengths was launched. As the data came in over its 11-month life-time. It confirmed one of the scientific predictions made in my third paper PULSE-SPHERES AND FLUXING DEAD ZONES. A UNIFICATION THEORY OF THOMAS YOUNG, AND EINSTIEN'S LIGHT THEORIES.

That was the first time someone else's experiment proved one of my predictions, while the last occurred in 1998. Two teams of Astronomers, one here in the United States, and one in Australia confirmed my prediction made in my 1997 paper (my second Peer Review Journal accepted paper), STRETCHED AND NEAR ZERO SPACE-TIME, A NEW THEORY OF GRAVITATION AND BLACK HOLE DYNAMICS. In it, I predicted and explained why the universe must be expanding at an accelerating rate of speed and not at Hubble's Constant. I not only predicted this but also explained why Black Holes went Bang!

13

I still don't think I am a genius though; I'm more convinced that everyone else is just that dumb. Now, I have come up with some brilliant ideas, I know that, but most of my work was based upon simple patterns that other people, far smarter than me, missed. For example:

To explain why Black Holes go bang after being stable for billions of years, I applied the *Law of Diminishing Returns* to gravitational phenomena.

We know that gravity diminishes over distance, and is a force of nature. It was only logical that it also diminished with increasing mass. *The Law of Diminishing Returns* and The *Law of Negative Returns* applies to everything, so it MUST apply to gravitation as well. That is simple!

To explain why the Universe was expanding at an accelerating rate of speed, I simply redefined the Black Hole, putting it into a state similar to a hypothetical Super-Compressed Boise-Einstein Condensate in Near Zero Space-Time rather than in Einstein's Zero Space-time.

Einstein held that there was no time in the Black Hole, but that means there cannot be any change in the Black Hole, as there is no time in which a change can happen. Near Zero Space-Time in the Black Hole

14

allows for a change in the Black Hole as there is time in which a change can happen, yet it allows change to happen so slowly that it explains why Black Holes remain stable for so long. We know the Black Hole changed, there was a Big Bang, so there MUST be Time in the Black Hole... Simple? Yes, simple. That is NOT genius.

"Where does the power of universal expansion come from?" you ask. Einstein did not proffer that answer! But again, it was simple logic.

To explain this, I redefined the essential nature of quanta in the Black Hole. Quanta resonate and interact with other Quanta naturally, but at Near Zero Space-Time in a Super Compressed Boise-Einstein Condensate, they are crammed together so tightly that they cannot resonate as individual Quanta, they are trying to do so...like people crammed into a room trying to get some elbow room, pushing outward against one-another. The quanta are likewise, pushing outward against each other, and against the compression of gravitation. In this situation, they become not quanta, but quantum potentials that only act as quantum springs pushing outward.

How does this explain the acceleration? Look at it, see it in your head...got it? Okay, now apply Galilei Galileo's laws of Force and Motion, that he developed in the 16th Century. All Motion is the result of a Net Force, not of a force, but of a Net Force. In the system, something must be moving against another thing to have a force, but to have a Net Force, a force must be greater than any other force acting upon the system. Look at the Black Hole in your mind, it's a ball of Super Compressed Quanta all of the quanta pushing outward in every direction trying to resonate. Now put a dot in the very middle of the Black Hole, and around that dot, draw concentric circles until you have encircled the very outer area of the Black Hole. Imagine these concentric circles being layers of compressed quanta. Each layer of quanta is pushing outward off the inner layer of quanta. Without gravitation, they would spring off of the layer of quanta below them and rocket out into space.

As the Black Hole sucks more and more mass into itself. The Expansive force of all that quanta gets larger. To constrain this larger outward force gravitation gets stronger too. However, eventually, the mass reaches the point of Diminishing Returns. When gravity entraps one more unit of mass, instead of gravity increasing one unit in strength, gravity increases slightly less than one unit. With the next unit of mass, the gravitational force grows but by slightly less than it did for the unit before it. As each unit of mass is entrapped by the Black Hole the increase in the gravitational compression grows by less and less. The outward expansive force of all those quanta is getting greater relative to the gravitational compression.

Eventually, the quanta start to get the better of the gravitational compression and the Black Hole starts to get bigger. As the Black Hole expands in size, the gravitational focus lessens. With this the Event Horizon shrinks moving closer and closer to the expanding quanta of the Black Hole. When the Expanding Black Hole crosses the Event Horizon, we get the Big Bang.

Now back to Galileo's laws of Force and Motion.
The First Layer pushes off of the second layer, the second layer pushes off of the third, the third layer pushes off of the fourth layer, the fourth layer pushes off of the fifth layer, etc. Nothing pushes back on the first layer of quanta, so nothing impedes its outward movement except

gravitation. The First layer pushes back against the second layer, slightly impeding the second layer and getting a boost from it. The Second layer and first layer both push back against the third layer and so impede it slightly more than the second layer is being impeded, and so get a boost from the third layer too. Each layer, as we go deeper into the black hole is impeded by one more layer of particles pushing backward and so, each layer has slightly less kinetic energy than the next outer layer. Thus, when these quanta escape the Black Hole at the Big Bang, each layer will be moving outward with a differing speed as they have different kinetic energies.

The innermost layers push outward with a force equal to the inward force of the outer layers pushing back. This is the Black Hole Remnant; these layers will not leave the Black Hole. The layers just outside of them will go so slowly outward, they will only make it inches outside the Black Hole then fall back into the Black Hole. The next layer will go a bit further, then it too will fall back into the Black Hole.

Each layer has greater kinetic energy than the inner layers so go out farther. Eventually, a layer will have so much Kinetic energy it will escape the gravitation and never fall back into the Black Hole. These quanta make up our universe and us. The layers outside from this layer of quanta have more kinetic energy, and so they are moving so fast that gravity is diminishing at a greater rate than the quanta's kinetic energy is diminishing. Thus, their energy over distance is growing relative to gravitation, and so, they are increasing in speed as they move away from the Black Hole. Thus, the universe must be expanding at an increasing rate of speed.

Now, isn't that SIMPLE? It is, it is stupidly simple. So why didn't anyone see it before me? I'm not a genius!

Then there is my discovery of what causes Inertia, also stupidly simple.

Acceleration is a transfer of kinetic energy from one body into another, which cannot take place instantaneously, it takes time to transfer energy. One half of the energetic pressure can transfer over, the other half can't, so we feel an equal resistance to any energy transfer, just like Isaac Newton said: *An equal and opposite reaction*. That lag in energetic transfer from one body to another is Inertia.

19

I do admit, seeing that this transfer explains the Twin Paradox was a brilliant leap, but was it genius? Let's look at the experiment run by Einstein.

One twin remains on the earth while the other is on a rocket ship traveling at .9c (9/10s the Speed of Light in a vacuum). Einstein saw that the rocketing twin would age slower than his twin on the earth. He incorrectly believed that this was due to a Warped Space-Time phenomenon. I realized that it was due to a change in the rocketing twin's inertial mass instead.

Remember, Kinetic Energy obeys Einstein's MASS ENERGY EQUIVALENCE formula ($E=mc^2$), which can also be written as ($m=E/c^2$).

As the Twin is accelerated by the rocket ship to .9c, Kinetic Energy is being pushed into his body by the spaceship's engines increasing his Kinetic Energy. When his Energy increases his Mass must also increase as mass and energy are *equivalent*! Einstein missed this for some reason that is totally unexplainable by me as it is so SIMPLE.

At 9/10s the speed of light in a vacuum, the Rocketing twin's body has more energy in it than the entire Himalayan Mountain range does. All that mass is squeezed into the volume of a 170-pound man's body. With his body so heavy the atoms of his body resonated slower. Just like a truck with a load of gravel in its bed that can't accelerate as fast as when its bed is empty. The same applies to the Twin at 9/10s the speed of light in a vacuum. As his mass increases his atom's resonance frequency slows down, or Red-Sifts, as we say in science.

"Time" is defined in science as the direction and the *rate of the flow* of Entropy. Entropy always increases, and that increase in the Entropic state is what we call Time. The direction of time cannot be altered, but, as Einstein made clear, the rate of the *flow of time* is relative to the Observer.

When you put meat into the freezer it lasts a lot longer because by cooling it down the atoms in the meat *vibrate slower*, so for the frozen meat, the rate of the *flow of entropy* slows down too, and so the meat ages slower. The same thing happens to the Twin, but not by cooling which removes kinetic energy from the meat, but by increasing the energetic mass of the twin's body through acceleration.

In the Space ship, in which everything is being accelerated at the same rate of speed as the Twin, the atoms of all the mass in and of the space ship are increasing in mass at the same rate as the Twin. This means it suffers from the same red-shift. Ergo, the chronometers on the space ship slow along with the Twin's atoms at the same rate showing no apparent change in the flow of time on the ship.

On earth, the other Twin ages normally while the Rocketing Twin ages much slower. Not, as Einstein said, due to warped space at .9c, but because of a simple mass/energy increase in the body of the rocketing twin's body.

Simple yes? Yes, it is. So, why was I the only one to figure it out?

Even Einstein missed it and he discovered Mass-Energy Equivalence! How could he have missed it? How could everyone in science have missed things that were so simple?

Now tell me, did you understand the ideas in my two papers? Are you an Einstein?

If you understand them and you are not an Einstein, then how can their writing have been the work of a genius when it is so easy to understand?

SIMPLICITY IS NOT SO SIMPLE

Simplicity: noun,

From the Latin *simplex* or *simplicitas*.

     1. The quality or condition of being easy to understand or do.

     2. To be free from extravagance, luxury, and complexity."

That seems to me to be a pretty straight forward definition of Simplicity,

don't you think?

Not complex, and easy to understand.

A Simpleton is a person who only has the ability to understand simple

things, like gravitational physics and quantum physics, right? Right,

because basically, the rules are simple.

     Einstein didn't like Quantum Physics because mankind had not

yet developed an understanding of the laws that governed Quantum

Particles. Every Physicist today will tell you; *Einstein would have no*

problem with Quantum Physics if he knew the rules that we know today.

They are simple and easy to understand rules, really, they are! But,

understanding even the simple things in nature, like Time, is also

Relative.

     It is completely illogical but it seems, only a genius can see

simplicity!

That makes no sense to me, a simpleton should see simplicity, but they don't.

The human mind loves complexity, and it looks for complexity everywhere, to the point of error. The mind loves complexity so much, people don't look for the simple answers, nor do they believe them. Only the genius looks for simplicity.

Yet one of the most important Rules in Science is Occam's Razor.

Occam's Razor was proffered by William of Occam was born in 1285, he went on to become one of the most important intellectuals of the 14th Century. He was a leader of intellectualism in Europe that would give birth to the Renaissance and the Age of Enlightenment. He helped to change how we thought, how we approached and measured Nature, and in doing so, helped to birth a Scientific Revolution. His most important rule was his Razor which tells us:

> *"The Simplest explanation for any observed phenomena is the Best explanation, until the simplest explanation can be proven to be wrong."*

Time and again Occam's Razor has been proven to be correct. While pompous intellectuals invent convoluted explanations for natural phenomena to prove how smart they are, in time they always fall to the simple explanation. Nature, is simple. The rules are simple. Thus, God is the Genius!

SO WHY ME?

Let me answer your first question:

"Why is a Peer Review Journal author in Physics writing a book refuting Man-Made Carbon Dioxide based Global Warming?"

I'm not a Climatologist, so what could I possibly offer a discussion of Carbon dioxide global warming?

I'm not a Climatologist, that is true, but that is okay, because when we are talking about Carbon Dioxide caused global warming, we are not talking about Climatology, we are talking about Physics. Why? Because the Sun's electro-magnetic particles, the Photons, are Quantum Particles that obey the Laws of Quantum Mechanics, that's Physics. How the cosmos interacts with the earth and affects it, is Physics. The rain, the wind, the ocean currents, you... it is all Physics. Physics is the study of the Natural Laws that control the universe and everything in it. Physics is the answer to "Why?"

It is impossible to understand how the sun warms the earth; or how molecules react when a photon hits them; or what happens when one does, without a basic education in Quantum Physics. These Climatologists pushing AGW, and all the other non-Physicists who claim Carbon Dioxide in the atmosphere is warming the earth, may be well meaning, but they are out of their field. None of them are educated in Quantum Physics, and so, they are getting it wrong. It is impossible for Greenhouse gases to warm the planet, or to cause a runaway greenhouse effect. These violate basic Natura Laws, that I will cover here in this work.

Frankly, I'm insulted that these people, the Climatologists, and all the others, with no education in my field, nor significant accomplishments in my field, are insinuating that I'm an idiot for studying Physics, when all you needed to do to understand Physics was to study Climatology.

And don't get me started on the Stars and Starlets of Hollywood, they are all idiots. They aren't out of their fields, they have no field, they are out of their minds. Not one movie star nor TV star has the qualifications to speak on the subject of Physics. Further, any scientist who speaks outside of their field is probably an idiot as well. If I went shooting off my mouth about their field of study with no education in it, they would think the same of me "Idiot!".

You NEED to have an education in a scientific field to understand it.

Let me ALSO say this… No Physicist anywhere on this planet supports Carbon Dioxide warming. It violates basic laws of Physics, like the STEPHAN-BOLTZMANN LAW, which is the law that actually controls the "Greenhouse effect".

In addition, the whole idea is based upon speculation drawn in the early 19th Century and never proven by any scientific testing. These ideas are from more than a century before we had the technology to measure what they were talking about.

When I think of all the tools we have today at our disposal in Physics, Astronomy, and Medicine. The 19th Century is a literally Dark Age in comparison. At any second of any day, I can log onto NASA's web sites and get real-time data from its solar satellites orbiting above the earth's atmosphere, or from its ground stations. No wonder they got things wrong in the 19th century, they couldn't test anything, and they didn't have the data we do today.

Before we get into the real-time data and the Natural Laws of Physics that prove man made carbon dioxide warming is impossible, let's go back into history and go over what they got wrong in the 19th

Century.

NATURAL PHILOSOPHY VERSUS SCIENCE

Natural Philosophy or "The Philosophy of Nature" was what people called the Study of Natural Laws before the advent of modern science, and it is considered to be the "precursor" of modern experiment-based natural science.

The name *Natural Philosophy* is much more appropriate than most people realize.

Science is the Formalized, Experimental Examination of phenomena, made in accordance with the Scientific Method. Meaning we test our ideas!

This definition limits what we call Science and thus, establishes what we call Faith.

Today, areas of Physics, like String Theory and Quantum Physics, are at or are nearing the threshold of Science. We are discussing phenomena that we may NEVER be able to test! Never be able to confirm scientifically. All we will have is our Faith that our numbers are correct, but Faith is NOT Science, it is Religion.

This would come as no surprise to Einstein who declared that his Belief in God was his:

"Awe of the Universe".

Einstein understood that at the central core of Mankind's knowledge of the Universe would always be Faith. We would never know it all, our technology would eventually fail us due to its limitations. Thus, Einstein knew, there was always going to be room for God in Science, because there was always going to be that great mystery of

"Where did it all come from, and How?

I have no problem with God in Science, we still have those big mysteries to face. Luckily for us, today we have far fewer mysteries to face than they did in the days of the Natural Philosophers. Before Max Plank created Quantum Physics, and modern science created the machines that allow us to measure nature as never before. Back when Marie Curie was killing herself with radiation because no one knew how deadly it was. These people were more Philosophers than scientists because they could not test all of their ideas. Science, thanks to their work, was being born, but it didn't fully exist yet.

By the end of the 19th Century Svante Arrhenius will publish his often-misquoted work:

ON THE INFLUENCE OF CARBONIC ACID IN THE AIR UPON THE TEMPERATURE OF THE GROUND. [April 1896, THE PHILOSOPHICAL MAGAZINE AND JOURNAL OF SCIENCE, Series 5, Volume 41, pages 237-276].

This work is both a presentation of his work and a summation of the work of the Natural Philosopher's during the 19th Century. Work he knows, in part, is incorrect because of his own work and the modern work of others. Yet he also furthers ideas that we know today are wrong because even in 1896 they still could not be tested, so false beliefs were held to be truth.

The work's most important offering to us is that he makes the first actual scientifically valid mathematical examination of the effect of Carbon dioxide, which he calls "Carbonic acid", and of water vapor upon the Sun's rays which warm the earth. His calculations show that adding Cardon Dioxide gas to the atmosphere will not warm the earth but rather cool it, and he calculates the actual Logarithm by which it will occur and shows that both Water vapor and carbon dioxide will cool the earth by preventing the Sun's heat rays from transacting the atmosphere and reaching the earth's surface.

THE HISTORY OF THE GREENHOUSE EFFECT

Ancient people knew that the Sun both illuminated and warmed the earth. They also knew that fire did the same thing. Thus, they concluded the Sun must be a fire. However, they thought that it was the visible light emission of the Sun and fire that illuminated the world and warmed them. They did not now about invisible Infrared Radiation (IR), nor did they even understand the scientific concept of radiation itself. However, they knew that the rocks around their fire pit became hot, but did not give off visible light as the fire did. This was ancient proof of heat without illumination, ancient proof that heat and light were different.

1800 a.d., Sir William Herschel Discovers Infrared.

It was known to Herschel, thanks to Sir Isaac Newton's work OPTICS, that the Sun's light was actually made up of colored light which could be viewed with the use of a glass prism. As it was incorrectly believed at that time that the visible light warmed the earth Herschel set out to measure how much heat was in each color of light that made up white light.

First, he went out in the direct sunlight and used a mercury thermometer to measure how much heat was being sent to the earth's surface by the Sun that day.

Then, he went inside where the south-facing window of his parlor had been boarded up, except for a small slit wherein which Herschel place a Glass prism. The Sun's rays went into the prism, through it, and out the other side casting a rainbow on a white placard on a table in front of the boarded-up window.

Herschel then placed a mercury thermometer in each of the colored lights measuring the heat in each and compared that measurement to a second thermometer in the room which told him the room's ambient temperature. When he added up all the heat in the visible light, he was befuddled. Nearly all the Sun's heat...was missing!

"Where had the heat gone?" he wondered.

If the visible light did not hold the missing heat, then Herschel understood his original thesis must have been wrong. There must be an invisible component of the Sun's radiation and it must possess the heat of the Sun. Herschel then placed the thermometer just outside of the red light where no visible light could be detected, and there he found it, the Sun's heat. Herschel had discovered invisible radiation. Today we call this Infrared Radiation (IR).

37

Jean-Baptiste-Joseph Fourier, The Father of the Greenhouse Theory.

BARON JEAN BAPTISTE JOSEPH FOURIER

(1768-1830)

Skip forward 27 years. Baron Jean-Baptiste-Joseph Fourier, Frances' Greatest Mathematician, is working on a puzzle. His calculations tell him that the earth is warmer than it should be if there was only the direct sunlight warming it. He decided that the atmosphere must be acting like an insulating blanket keeping the sun's heat in making the earth warmer than mere sunlight could. His calculations told him that the earth was 59F degrees (15C) warmer than it should be. If you have ever wondered where people get the 59F degrees attributed to the Greenhouse effect you keep hearing about, this is it, and it is way off. But Fourier in 1827 had no means to measure the sun's rays, let alone did he know how much of each type of ray were reaching the earth. So he guessed...wrong. Fourier was correct, there is heat being sent back by the atmosphere that we today call *"Backscatter Radiation"*. The atmosphere does act as an insulating blanket. But it actually blocks out most of the Sun's heat, letting very little in. As the atmosphere blocks

out far more than it blocks in, the so called "greenhouse gases, cool the

earth by over 180F degrees, creating a Net Cooling effect, not a

warming one. We will go over the Physics of that later in the book.

Right now, I just want you to know where the 59F degrees came from.

FOURIER DISCOVERS HIS GREENHOUSE EFFECT...NOT!

Now let's skip forward 2 years to 1829. Fourier is now working on

another puzzle, but this one is of his own error's making.

Fourier was, in all respects, a genius mathematician who did

revolutionary work studying heat. But he got a wrong idea fixated into

his head. He believed that glass blocked infrared radiation!!! Yes,

that's right, the glass in the glass prism Herschel used to discover

Infrared Radiation with…Fourier held actually blocked IR radiation

rather than allowed it to pass through the glass! How Fourier got this

wrong idea no one knows, but like many wrong ideas, it caught on.

Obviously, Fourier was wrong. If glass did block IR then no IR

would have entered into Herschel's parlor through the glass prism. So,

Herschel would never have discovered IR to begin with. This simple

thought never seemed to enter into Fourier's mind, nor into other men's minds. This idea that glass blocked IR radiation led Fourier to ask himself:

> *"If IR cannot pass through the glass of a Greenhouse why does it get hot inside?"*

The Visible light, Fourier decided can pass through the glass and in striking the interior the visible light must be warming the interior. It is then turned into IR and then emitted out. The new IR cannot pass through the glass and so is trapped inside warming the interior. Fourier believed that he had discovered what he called *"The Greenhouse Effect"*.

It would be 84 years before Prof. Robert Wilson Wood scientifically put Fourier's idea of a Greenhouse Effect to the test and prove him wrong. By then the idea had become so mainstream that it didn't matter, people still believed it to be true.

The idea was idiotic, if the Sun's visible light was turning into invisible IR, then how could we see anything? If everything is giving off invisible light then it should be invisible? The inside of a greenhouse should be dark.

Today NASA satellites measure all radiation, light, heat, Microwaves, and radio coming and going, and the heat being given off by the earth is equal in amplitude and wavelength to that which is coming in as this NASA chart shows:

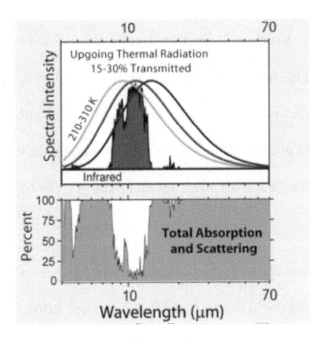

The outgoing IR (in blue), is identical to the IR striking the Earth's surface (white against gray). This would be impossible if visible were turning into IR as the average visible light photon has a thousand times more energy than the average IR photon. Given the energy difference between visible and IR, at least a thousand times more heat should be leaving the earth than entering if visible is turning into IR, yet to NASA satellites, this extra visible light turned into invisible IR heat is nowhere to be found.

The entire idea of Carbon Dioxide warming is based upon this idiotic idea that Visible radiation turns into heat, creating an imbalance in entering heat and exiting heat.

The obvious proof that visible light is NOT turning into heat can be found on any overcast day.

On an overcast day visible light is still coming in, clouds block out IR heat, not visible light. Yet it gets cold on an overcast day. If Visible light was the factor warming the earth as Fourier thought, then the earth should get warm on an overcast day, not cooler.

Clouds are 100,000 times more insulating than CO^2. So, with the atmosphere keeping in 100,000 times more heat, on an overcast day, it should be warmer if visible light is turning into IR, but it gets colder, because...the clouds block out 100,000 times more IR from the Sun, and it is the IR that warms the earth not the visible light.

The Global Warming crowd does not deny that the Carbon Dioxide blocks out some Sun heat, but they incorrectly claim that the IR portion of the Sun's radiation is tiny, like Fourier did, and the visible portion is huge. Just as Fourier claimed in 1829, even though he had no way to measure this. Since Fourier believed that more heat was leaving due to visible light truing into IR than was entering, he believed in a Greenhouse Effect powered by visible light turning into IR heat. Today we know this is not happening, outgoing heat is equal to incoming. Visible light is NOT turning into heat and creating a greenhouse effect.

NASA satellites tell us that IR makes up 52% of the Sun's radiant emissions, while the visible portion only makes up 42%, with UV making up the other 10%.

It is not just lay people who still hold to Fourier's outdated beliefs. I was stunned when, in e-mail discussions with noted and award-winning Climatologist Dr. Roy Spencer (an opponent of Man-Made Carbon Dioxide global warming), and noted Physicist Dr. Richard Mueller (a pro-Greenhouse Warming proponent) they each repeated Fourier's belief that the IR was miniscule in size compared to the visible proton of the sun's radiation. I'd asked them to give me some feedback on an article I'd written on the Stephan-Boltzmann Law's application to the Greenhouse Effect. In our discussions, each man repeated Fourier's erroneous belief that the Sun's IR was a tiny portion of its emissions and so the blocking of it by the atmosphere didn't cause any significant change. Thus, both men held the Stephan-Boltzmann Law had little effect upon the discussion of added Carbon Dioxide to the atmosphere!

Since then, I've noticed many Degree holders in the best educational institutions like Harvard, Yale, MIT Caltech, Berkeley, etc. are repeating ideas from the early 19th century that were scientifically proven to be wrong long ago.

Let me repeat, Fourier's beliefs, were and still are, totally without scientific support, and have been proven wrong scientifically. It was impossible for Fourier to test his idea at that time because he did not have the technology that would have allowed him to do so. This erroneous belief of Fourier's that visible light is what warms the earth, and that IR is minuscule and unimportant is the primary argument of

Greenhouse Gas Warming theory today. It was wrong in 1829 and it is wrong now.

NASA satellite data tells us that the IR heat being given off by the earth's surface is equal to what it is receiving from the sun, there is nothing extra leaving, and the fact that we can see means that visible light is not turning into invisible IR radiation.

ON THE INFLUENCE OF CARBONIC ACID IN THE AIR UPON THE TEMPERATURE OF THE GROUND

On the Influence of Carbonic Acid in the Air upon the Temperature of the Ground, By Svante Arrhenius, <u>Philosophical Magazine and Journal of Science,</u> Series 5, Volume 41, April 1896, pages 237-276.

It is widely, and incorrectly held that Svante Arrhenius won the Nobel Prize in 1903 for the above work on Carbon Dioxide and Water Vapor's effects on the transmission of heat. He in fact won the Nobel Prize in Chemistry:

> *"In recognition of the extraordinary services he has rendered to the advancement of chemistry by his electrolytic theory of dissociation."*

It is also widely and incorrectly said, that in the above work he claimed that CO^2 was warming the earth. In fact, he proved the opposite.

Given the lack of technology in the 19th Century, even at its end when Arrhenius lived, it was impossible for much of what people believed about the laws that govern the sun and its radiation to be tested and confirmed. The first satellite to tell us what kinds of radiation the sun gave off was not launched by NASA until 1979. As such, Arrhenius, and others held to many things that we today know were wrong. In his paper, Arrhenius re-states many of these errors, such as the belief that the IR portion of the Sun's radiation is the *"'lessor part"* and the visible-light portion is *"the greater part"*. The IR portion we know today is 2800 times larger than the visible part of the Sun's spectrum and constitutes 52% of the Sun's total radiant energy. Not only does Arrhenius not say CO2 is warming the Earth in this work, he states the opposite writing on page 240:

> *"The intensity of radiation for any group of rays should **always diminish** with **increasing quantity** of aqueous vapour or carbonic acid traversed."*

While on page 244 of the work he gives his exacting calculations writing:

> "*If a ray of heat, corresponding to the angle of deviation 39 degrees 45 minutes passes through the unit of carbonic acid, it **decreases in intensity** in the proportion 1: 0.934 (log= **-0.0296**), the corresponding value for the unit of water-vapour is 1: 0.775 (log= **-0.11105**).*"

Oddly enough, while he notes the much greater effect water vapor has upon heat compared to the lessor effect of Carbon dioxide, he writes on page 239:

> "*The researches of Paschen show that these gases are **both very effective**, so that probably sometimes the one, sometimes the other, may have the greater effect according to the circumstances.*"

Today satellite technology allows us to know that water vapor is responsible for 66% of the heat absorption in our atmosphere. His paper thus proved Pachen's work was incorrect, yet he seems to have missed this. Yet he calls Langley's work absurd writing on page 245:

49

*"Below A is written the relative value of the intensity of radiation for a given kind of ray in the moonlight after it has traversed K=1 and W-0.3. In some cases, the calculation gives **positive values** for log x or y. As this is **a physical absurdity** (it would signify that the ray should be **strengthened** by its passage through the absorbing gas.)."*

What Arrhenius' paper establishes is that the Sun's heat is being absorbed by the "aqueous vapour" at a log of -0.11105, and absorbed by "carbonic acid" in the atmosphere at a log of -0.0296, and so is cooling the Earth by blocking out the Sun's heat. An effect which we can easily experience on cloudy days. Clouds block out the sun's heat and cools not the earth.

THE HISTORY OF CARBON DIOXIDE AND EARTH HEAT

Foraminifera, Nature's Thermometer

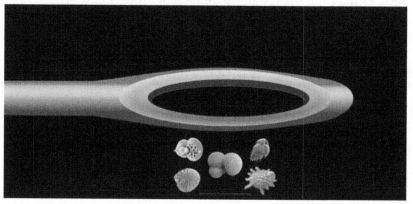

The chart below shows the past Six-hundred million years of earth heat and its Carbon Dioxide levels. This data, like all such ancient data, is based upon data derived from Foraminifera. Forams are single-celled, water living creatures that are the size of the period at the end of this sentence. "." They were evolved 650,000,000 years ago and are found, after 600,000,000 years ago, in all aqueous sediment on the planet. Forams have evolved to live in hot water, cold water, saltwater, briny water, brackish water, muddy water, and freshwater. They have conquered the Earth's watered areas. More importantly, they make shells of calcium-carbonate that survive their deaths as fossils. These fossil shells remain in the depths until scientists dig them up and examine their chemical make-up for clues to the earth's Climatologic past.

Paleo-climatology is the study of ancient climate and the Foram is its greatest tool. Since the Foram takes its shell building material from the water itself, the Foram's shell is a historical record of the chemical composition of the water at that time. Today, we know that the chemical composition of water changes with its temperature. However, Forams can not only tell us the temperature of past waters, but also the chemical composition of past atmospheres, and how much ice there was on the Earth at that time.

Thanks to Foram shells we have exacting climate and atmospheric data for the past 600,000,000 years. This should come as quite a surprise to you. You have been reading and listening to Liberal lies and talk about AGW for decades and yet none have ever mentioned Forams, this is the first time you have probably ever heard the word uttered! Former Vice President Gore has made two fraudulent films about AGW, and he never mentioned Forams, nor the 600,000,000 years of temperature and carbon-dioxide data they provide. The reason for this silence is shown in the chart.

600,000,000 YEARS OF FORAM DATA.

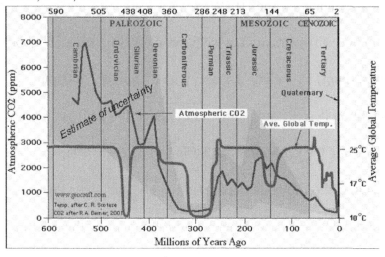

The chart shows Foram data for the past 600,000,000 years. Specifically, it shows the Earth's temperature and the amount of Carbon-dioxide in the atmosphere over that time.

Remember in its 5th report the United Nations International Panel on Climate Change (IPCC), stated that Carbon-dioxide in the air is responsible for "**100%**" of the temperature change that the Earth has seen since the end of the Little Ice Age in 1855. According to the IPCC when CO^2 levels rise, earth's temperature rises. When CO^2 levels fall, Earth's temperature falls. Thus, the IPCC hold that CO^2 is the dominant Greenhouse Gas of them all. Responsible for all current warming and that Earth temperatures move in pace with CO^2.

Looking at the Foram data chart for temperature and Carbon-dioxide levels over the past 600,000,000 years, on the far left we can see for the first 300,000,000 years of data we have, the Earth averaged 4,000 parts per million (ppm) of Carbon-dioxide, with a high of 7,000ppm during the Cambrian Period, and a low of just under 400 ppm (similar to today's levels) during the Carboniferous Period. There was a long gradual decline in carbon-dioxide in the atmosphere reaching a low during the Carboniferous Period. However, during much of this decline, the Earth sees no change in its temperature.

Over the most recent 300,000,000 years, the Earth averaged 1700ppm of carbon-dioxide. Eventually falling to today's levels at just under 400ppm.

Only two times in the past 600,000,000 years has the Earth seen temperatures as low as today. Only once has the Earth seen carbon dioxide levels as low as today!

We can see at the end of the Ordovician Period, with carbon-dioxide at more than 4,000ppm, temperatures plummeted down to below today's temperatures while carbon-dioxide levels were **rising**. Clearly no relationship between carbon-dioxide in the air and temperatures there.

This cold snap ended at the start of the Silurian Period, however, as temperatures skyrocketed back to their past norm, carbon-dioxide levels were **falling sharply** down to 3,000ppm before jumping back up to 4,000 ppm. No relationship between temperature and carbon-dioxide levels there either.

From the Mid-Devonian Period both temperature and carbon-dioxide levels seem to be falling apace of each other, then they break away with temperatures holding steady for over 30,000,000 years while Carbon-dioxide continues its steady drop to today's low levels. After a 30,000,000 year pause in temperature movement, temperatures rapidly fall to below today's temperatures and remain there for another 30,000,000 years.

Just after the start of the Permian Period temperatures start to rise then quickly accelerate at the Mid-Permian reaching, at the Permian/Triassic border, the second-highest temperature spike of the past 600,000,000 years. However, carbon-dioxide levels remained below 2,000ppm, half that of the historical average for carbon-dioxide.

Yet this Permian/Triassic spike in temperatures lasts for millions of years.

Temperatures fall back to the average once into the Triassic Period and hold steady for over 50,000,000 years, while carbon-dioxide falls down to only 1,200ppm, then rises back up to 2,000 ppm. However, as carbon-dioxide levels are rising throughout the Mid-Jurassic Period, the Earth sees its third significant **drop in temperature** as the Earth enters the Cretaceous. After some 10,000,000 years at this low in temperature, temperatures start to rise, but carbon-dioxide levels are **falling**!

Again, with carbon-dioxide levels falling from over 2,000ppm to 880ppm, the Earth's temperature holds steady at its historic average for over 50,000,000 years. Millions of years into the Tertiary Period, with carbon-dioxide levels falling, the Earth sees its highest temperatures, followed by a ragged 50,000,000-year decline in temperature down to today's levels, with the Earth seeing low carbon-dioxide levels for only the second time in the past 600,000,000 years.

If there were any link between high Earth temperatures and high carbon-dioxide levels as claimed by the IPCC, then the highest temperatures would have occurred during the Cambrian Period when carbon dioxide levels were always over 5,000ppm, not during low levels of carbon-dioxide as we see occurring in the Permian/Triassic border and in the Tertiary Period. The Earth has been this cold only three times in the past 600,000,000 years and, in two of these cool periods there was low carbon-dioxide levels but high carbon-dioxide levels in the first low-temperature period at the end of the Ordovician Period where carbon-dioxide was over 4,000ppm. Conversely, the highest Earth temperatures are seen during historically low carbon-dioxide level periods, not during the highest carbon-dioxide levels.

The Forma data clearly shows that there is no identifiable relationship between carbon-dioxide in the atmosphere and Earth temperatures over the past 600,000,000 years to support the claim of a carbon-dioxide link with higher temperatures. The Earth has held its average temperature with 7,000 ppm carbon-dioxide and with only 500ppm. Obviously, if carbon-dioxide was as powerful as the IPCC says it is, to be responsible for 100% of the temperature change we are seeing since the end of the Little Ice Age, then we should see a clear relationship throughout Earth's climatologic history between carbon-dioxide levels and Earth temperature, yet we see none. Earth temperatures historically moved with indifference to, or in opposition to carbon-dioxide levels throughout much of the history that we have data for. With no historical link to be seen over the past 600,000,000 years, it is unscientific and fraudulent to claim that there is a statistical link over the last 150 years. The Liberals are cherry-picking the last 150 years to make their point while ignoring the past 600,000,000 years that prove them wrong.

HOW SOLAR RADIATION CREATES HEAT

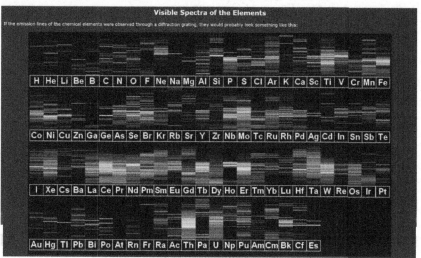

Visible Spectra of the Elements

If the emission lines of the chemical elements were observed through a diffraction grating, they would probably look something like this:

All atoms absorb some bands of solar radiation, even when an atom is bound inside a molecule. It is the individual atom in the molecule, not the molecule itself, that absorbs the photon. It is the individual atoms of Hydrogen or Oxygen in the water molecule that will individually absorb a photon of IR. Whether the water molecule is in its gaseous state (water vapor is microscopic water droplets), or in its liquid state as larger water droplets in a cloud, it doesn't matter, the atom will interact with a photon in the same way.

The chart above shows all of the elements in the Periodic Chart. As you can see, they all react with radiation, both visible as here presented and invisible. Astatine (symbol "At", bottom row, 7th from the left), only absorbs invisible radiation. While Thorium (Symbol "Th" bottom row 12th from the left), seem to absorb almost every visible color. Which type of radiation the atom absorbs is dependent upon the Photon's energy.

In High School you probably learned that only one of any three possibilities will occur when an atom meets up with a photon: Absorption, Reflection, or Transmission.

This is a shorthand that is not actually correct. All solar radiation will be absorbed, but if the photon does not match the atom, it will immediately spit it out via either Reflection (back out off the surface it hit), or, if the mass is transparent to that photon, it will be transmitted through the mass and out the back. If the photon matches up with the atom then after the initial absorption, it will be held a bit longer and interact with the mass.

An atom has three states that can be affected when it interacts with a photon. The State of Torsion and Rotation, The Energetic State, and the Vibrational State.

Microwave Interactions- States of Torsion and Rotation

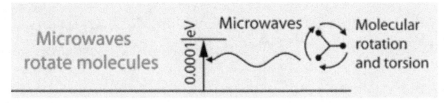

Microwave photons have energies between 0.00001 to 0.001eV

(electron volts). This is in the range of energies separating the quantum

states of molecular rotation and torsion. Interaction by mass, other than

metallic conductors, will cause the molecules to rotate and produce heat

as a result of that molecular motion. This is how your microwave oven

works. The entire mass of the atom is changed by the absorption of the

microwave photon creating noticable heat.

Infrared Interactions- Vibrational State.

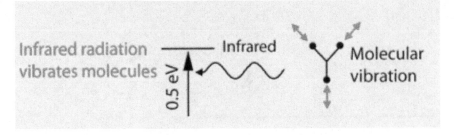

Infrared photons have energies in the range of 0.001 to 1.65eVs. This is the range of energies separating the quantum states of molecular vibration. When an Infrared photon is absorbed it changes the Vibrational State of the atom causing the atom to vibrate faster. This change in vibration is significant because what we refer to as heat is actually the rate of the atom's vibration. The faster an atom vibrates the more heat we say it has. The slower it vibrates the less heat it has. As the entire atom vibrates faster this creates noticeable heat. However, this is not true for all IR radiation.

Near-Infrared radiation has energies close to visible light. In fact, scientist even disagree where the Infrared begins. Some say 750 nanometers, others say 800 nanometers. Like visible red light, Near-IR creates no noticeable heat. Your TV remote control uses Near-IR to change the channels, yet you have never noticed any warmth from that Near-IR have you? At the other end of the IR spectrum is the Far-IR. This radiation creates a lot of heat, just like a fire.

Visible Light Interactions- The Energetic State.

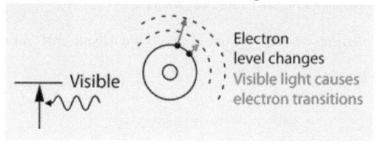

Visible light photons when absorbed by an atom changes its Energetic State.

This causes a single electron in a lower energy shell, orbiting around the nucleus of the atom, to jump to a higher energy level shell. Once at the higher energy level, the electron emits out the photon it absorbed and falls back to its resting energy level. Because only a single electron moves in the atom, there is very little movement and so no noticeable heat is created.

This is why the EASYBAKE OVEN now uses an electrical heating element in its oven. Incandescent light bulbs emit 90% of their energy as IR photons and only 10% as visible light. The incandescent light bulb, however, has been replaced with Florescent and LED light bulbs, that emit 90% of their energy as visible light and only 10% as IR. Since visible light causes so little movement in the atom it is unable to produce useful heat and so it cannot bake a mini cookie or cake in the EASYBAKE OVEN. So, the manufactures had to install an electric heating element similar to the one in an electric oven in order for their

product to work.

HOW THE ATMOSPHERE INSULATES THE EARTH

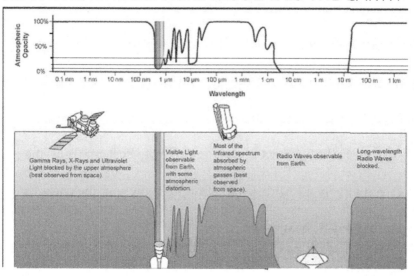

The chart above shows the Opacity of the Earth's atmosphere. It shows what wavebands of light are being blocked out by the earth's atmosphere.

On the left, you can see all the Gamma rays, and X-rays are blocked out along with 70% of the UV. On the far right, you can see all the Long-wave Radio waves are blocked out. While in the middle you can see nearly all the Far-IR heat is blocked out, along with most of the Microwaves. Only the AM and FM bands of radio waves get through without blocking. The rainbow of visible light shows that most of the visible light gets in. Only 6% of the green and blue light is blocked out. Yellow and Red have higher rates of blockage with almost 25% of the red being blocked out. To the right of the rainbow is the Near-IR. Most of the Near-IR is blocked with only a few bands getting past the

atmospheric blocking in any significant level. To the right of the Near-

IR is the Mid-IR. Only 12% of this is blocked out and so the earth

primarily receives all of its heat in the form of Mid-IR heat. With the

Earth receiving largely all Mid-IR radiation from the sun, the earth

primarily emits Mid-IR heat as well.

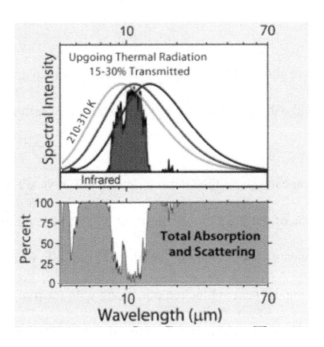

Returning to the earlier used NASA chart, we can see the dominance of the Mid-IR both incoming and outgoing. This chart also proves that visible light is not changing into IR as Fourier, and his "Greenhouse Effect" followers claim. The heat leaving the earth is the same as the heat that has entered through the atmosphere. A perfect balance of incoming and outgoing. If visible were turning into IR then there would be a great deal more heat leaving than entering.

So, how does the atmosphere block out the sun's heat and help keep in some of the earth's emitted heat?

Today, the popular phrase non-scientists use is "Greenhouse gas molecules", or GHGs. However, there is no such thing as a Greenhouse gas, as all gases cool the Earth far more than they warm it.
In science, these gas molecules are called Thermally Reactive Molecules, or ThRMs for short. ThRMs are molecules that absorb IR photons and the most powerful of them is the Water molecule.

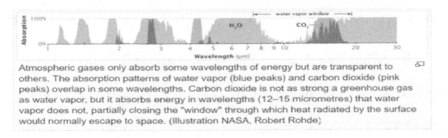

Atmospheric gases only absorb some wavelengths of energy but are transparent to others. The absorption patterns of water vapor (blue peaks) and carbon dioxide (pink peaks) overlap in some wavelengths. Carbon dioxide is not as strong a greenhouse gas as water vapor, but it absorbs energy in wavelengths (12–15 micrometres) that water vapor does not, partially closing the "window" through which heat radiated by the surface would normally escape to space. (Illustration NASA, Robert Rohde)

The chart above shows the absorption of Water molecules and Carbon Dioxide molecules. As the chart above shows most of the Mid-IR heat that reaches the earth is in the 10-micron range. This is due to the water molecule's poor absorption at this range. However, Carbon Dioxide is a strong absorber in the 12-15-micron range and so takes up some of the blocking of Mid-IR heat here. The 10-micron range is known as the Water Vapor Window. Most of the earth's heat comes from this area. In the previous section, you learned what happens when a Visible, IR, or Microwave photon is absorbed. Now we see what happens after it is expelled by the ThRMs.

Like most people you have probably played Pocket-billiards. The balls on the table move in predictable ways according to simple natural laws. If the cue ball hits a ball straight on, the Cue ball will transfer its energy and momentum into the ball it hits and that ball will continue traveling in the straight line the cue ball was traveling in. If the cue ball hits a ball off to one side, they will go off in opposite directions

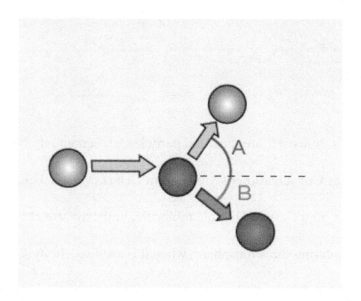

This predictability is made possible due to the *Law of Conservation of Energy,* and the *Law of Conservation of Momentum.* When the cue ball hits another ball, it will always move off in a direction relative to the cue balls momentum (direction). It will never move in the opposite direction. The cue ball transfers some, or all, of its energy and momentum into the ball it strikes so it must move in relation to the cue ball's energy and momentum.

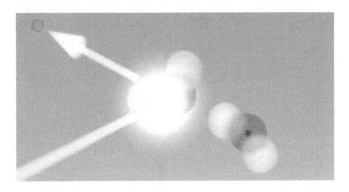

Photons, however, are quantum particles, and quanta do NOT obey the Law of Conservation of Momentum. Thus, when a ThRM water molecule, or a Carbon Dioxide molecule, in the air absorbs a photon of IR entering the atmosphere, when it is emitted out by the atom, the outward direction of the photon is random. This is due to the fact that when a photon is absorbed, it is destroyed completely, it no longer exists. When the atom emits the photon out, in that instant, a new photon is created. It is a new photon, not the old one. As a completely new photon, it has no past, no past energy and no past momentum. Thus, its new momentum is totally random. So, each atom emits out photons like a mini-sun, it sends them out in every direction.

Most of these photons absorbed by the earth's atmosphere are sent out back into space. They are emitted up, or out or to the sides, but some, 29.4% of them, are emitted downwards towards the earth. These photons land somewhere upon the surface of the earth warming that spot.

Likewise, when the earth radiates off the sun's heat it received, the IR photons are absorbed by the atmospheric ThRMs on their way out too. Since the number of ThRMs in the atmosphere is identical to incoming and outgoing IR photons, the same percentage is reflected downward towards the earth and so strike the earth for a second time, along with the new photons coming straight through the atmosphere.

So, the question to be asked is, how much energy is being sent back, and how much heat does it add to the incoming sun heat?

HOW MUCH HEAT IS BEING KEPT IN BY THE ATMOSPHERE?

NASA reports that 29.4% of the heat emitted off the earth is returned to it by the ThRMs in the atmosphere. This is equal to about $61 wm^2$ (watts per meter squared).

As NASA says, 70.6% of the Sun's heat is absorbed and set back out when it tries to cross the atmosphere and reach the earth's surface. This is done mainly by water vapor. This direct sun heat is absorbed and then radiated in every direction. 70.6% of the IR photons are emitted outwards to the sides and above, going back into space while 29.4% of the IR photons are emitted downwards towards the earth.

Likewise, 70.6% of the earth emitted heat heading out to space is absorbed by the ThRMs and emitted out into space, while 29.4% of these outgoing IR photons are absorbed by the ThRMs and sent back down to the earth returning to the surface it just left. This returning heat is called "Backscatter radiation".

Once the earth warms up during the daytime, at any time of the day, 29.4% of the heat you feel is backscatter radiation from the atmosphere. Thus, the R-value, or the insulating value of the earth's atmosphere to outgoing earth emitted heat is an R-value of 29.4%. While the R-value to incoming Sun heat is 70.6%, as it insulates the earth from that heat, keeping it from burning the earth's surface to a crispy 254 to 265F degrees at the equator. The ThRMs keep out 70.6%

of the Sun's total heat while keeping in only 29.4% of 29.4% of the

sun's heat that reaches the earth's surface. So, the backscatter radiation

is equal to **8.6436%** of the TSR, the Sun's total direct sunlight arriving

at the top of the earth's atmosphere.

Now, let's do the math?

On a day in which the sun's direct sun light reaching the earth's

surface is equal to 60F degrees. Once absorbed, the earth then emits that

60F degrees back out into space. But 29.4% of that 60F degrees of

outgoing heat is going to be absorbed by the atmosphere and returned to

the surface of the earth. So, 29.4% of 60 =17.64F degrees of backscatter

heat radiation being added to the Sun's 60F degrees, giving you a total

temperature of 77.64F degrees.

So, right now, with the amount of thermally reactive molecules in the

atmosphere when it is 77.64F degrees out, 17.64F degrees of that heat is

backscatter radiation.

If the sun's direct heat is 70F degrees, 29.4% of that exiting heat is absorbed and sent back as backscatter thermal radiation adding 20.58F

degrees to the day's heat for a temperature of 90.58F degrees!

Remember Fourier's claim that the backscatter radiation was

equal to 59F degrees? How much direct sunlight coming in must the

earth get to have 59F degrees of backscatter radiation?

In order for the backscatter heat radiation to be equal to Fourier's

59F estimate, the direct sunlight would have to be equal to 200.68F

degrees, as 29.4% of 200.68 is equal to 59F. According to the World

Meteorological Organization, the highest registered air temperature on

Earth was 134.1F (56.7 °C) in Furnace Creek Ranch, California, located

in the Death Valley desert in the United States, on 10 July 1913. That is

far less than the 200 plus degrees needed to create a 59F degree

backscatter radiation. But why was Fourier so far off? Again, he was a

Natural Philosopher, he did not test any of his theories scientifically.

Plus, he had no modern technology that could give him a real

understanding of what the sun and atmosphere were doing. Remember

he also thought that the visible light was what warmed the earth and that

the IR portion of the sun's emissions was insignificant. Today we have modern technology so, we know Fourier's calculation of the backscatter radiation was very wrong.

THE STEFAN-BOLTZMAN LAW

By the close of the 19th Century Science had correctly concluded that all matter warmed above absolute zero degrees Kelvin would radiate some thermal radiation in the infrared bands, but a formula to allow us to calculate how much thermal energy was being radiated was needed. In 1879 (fifty years after Fourier's work), Austrian physicist Josef Stefan provided it. He created the constant known today as the "Stefan Constant",

$$\sigma = \frac{2\pi^5 k_{\mathrm{B}}^4}{15 h^3 c^2} = \frac{\pi^2 k_{\mathrm{B}}^4}{60 \hbar^3 c^2} = 5.670373(21) \cdot 10^{-8} \ \mathrm{J\,m^{-2}\,s^{-1}\,K^{-4}}$$

expressed in Stefan's Law as:

$$j^\star = \sigma T^4.$$

Meaning that the amount of energy given off by a Black Body in joules of energy, is equal to Stefan's Constant (Sigma), times the Fourth Power of the Temperature of the black body (T^4). This tells us the amount of energy given off by a Black Body at any known temperature, but the earth is not a black body. Stefan's student, Ludwig Boltzmann, furthered Stefan's work to include non-Black Bodies by including their "emissivity" to give us the Stefan-Boltzmann Law:

$$j^\star = \varepsilon \sigma T^4.$$

This tells us the amount of energy a body will emit in joules of energy, is equal to Epsilon (its emissivity), times Sigma (Stefan's Constant), times the Fourth Power of the temperature.

The Stefan-Boltzmann Law's tells us that any change in temperature, up or down, will cause a much greater change in the amount of radiation given off by the mass. A 1% drop or rise in temperature will create a 4% change in emitted heat off the surface of the mass. This is why a Runaway Greenhouse Effect is impossible, as NASA says here:

http://earthobservatory.nasa.gov/Features/EnergyBalance/page6.php

"Why doesn't the natural greenhouse effect cause a runaway increase in surface temperature?

*Remember that the amount of the energy a surface radiates **always increases faster than its temperature rises** -- outgoing energy increases with the **fourth power of temperature**. As solar heating and "back radiation" from the atmosphere raises the surface temperature, the surface simultaneously releases an increasing amount of heat--equivalent to about **117 percent of incoming solar energy**."*

This rule also holds for a DROP in surface temperature. The amount of energy a surface radiates always decreases faster than its temperature decreases, at a rate equal to the fourth power of the drop in absolute temperature. How does this apply to the Greenhouse Effect (GHE) you ask?

It is the Greenhouse Effect.

The Greenhouse Effect is powered by the heat radiated **off of the Earth's surface**, not by atmospheric heat, only by the earth's emitted heat!

The ThRMs (Greenhouse gases), mostly water vapor at the bottom of the atmosphere absorbs some of this outgoing earth emitted heat and sends it back down to the surface, currently adding just over 61 watts per meter square of energy to the incoming Sun heat.

In order to increase the amount of backscatter heat the earth gets, you must increase the amount of heat the earth gives off. But the heat the earth gives off is totally dependent upon the heat it receives from the sun. If the Earth gets more heat from the sun THEN it will emit more heat out, increasing the amount of backscatter heat being sent back at the earth by the atmosphere's ThRMs. If the Earth receives less heat from the sun, then less heat will be sent out by the earth's surface, and so less heat will be backscattered back at the earth by the atmospheric ThRMs, decreasing the greenhouse effect.

This is why clouds cool the earth so effectively. Clouds block out a lot of Sun heat while letting the visible light through. With so much of the sun's heat being blocked out by the clouds the earth is a lot cooler on a cloudy day, so it gives off way less heat. There is so much less heat leaving the earth on heavily overcast days that the backscatter heat is almost insignificant.

How does this prove that warming via increased amounts of carbon dioxide is impossible?

All the ThRMs absorb incoming Sun heat and currently send 70.6% of it back out into space, greatly cooling the earth. The Moon has no atmosphere to absorb the Sun's radiation and its daytime temperature is 256 to 264F degrees Fahrenheit! Hot enough to boil water.

If we increase the amount of ThRMs in the atmosphere, carbon dioxide, methane or water vapor, that means that more incoming solar IR will be absorbed and reflected back out into space by the additional ThRMs, reducing the amount of sun heat that can make it to the surface, cooling the Earth more than the current level of ThRMs do now.

Just like with clouds on a cloudy day, less heat will reach the earth with more CO^2 molecules in the atmosphere to absorb incoming Sun heat photons. With less sun heat reaching the earth, the amount of heat the earth will radiate off will also be less, thus, there will be less

backscatter heat, reducing the earth's temperature twice. Less direct sun

heat, and less backscatter heat.

But remember with a small drop in temperature, we get a larger

drop in emitted radiation off the earth, and that means much less

backscatter radiation.

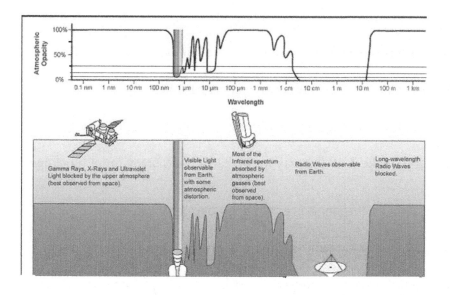

NASA provides this chart showing the Atmospheric Opacity to

incoming solar radiation. It shows the amount and type of Solar radiation that is being blocked by the earth's atmosphere now with the current level of ThRMs in the air.

Looking at that NASA chart showing the opacity of the Earth's atmosphere, you can see on the far left 100% of the Gamma and X-Rays are blocked out. 70% of the UV is blocked out. Between 6% and 11% of the visible light is blocked out. Over 62% of the Near-IR is blocked out, while only 12% of the Mid-IR (10 micrometers or 10-micron range IR radiation) is blocked out. Almost all the Far-IR is blocked out. Additionally, most of the Microwave energy is blocked out as well as the Low Radio waves energy. Only the AM and FM bands of Radio wave energy make it all the way to the Earth's surface without blockage.

However, this chart is based upon current levels of trace gases and water vapor in the atmosphere. Increase the levels of these trace gases, or water vapor and they will block out more of the radiation that currently makes it to the surface of the Earth cooling the Earth.

Let's do some more math?

Here is a web site with a free to use Stefan-Boltzmann calculator: http://calculator.tutorvista.com/stefan-boltzmann-law-calculator.html

You'll be able to run your own tests to see how much less or more heat you get when you change the absolute temperature of the Earth's surface. Let's start with a temperature of 100 degrees Kelvin. The earth's emissivity is 0.5, and use an area of 1 square meter to keep things easier to calculate.

Go to the web site and put an "x" in the "Radiation Energy" box. The "x" is what we are looking to find out.

Enter 100K in the Temperature box and 1 in the area box.

At this temperature, the calculator tells us that a 1 square meter of earth will emit 2.835watts of energy at 100K.

Now let's reduce the temperature 1% which is 1degree K. At 99 K we only get 2.72328968835 watts of energy being emitted, that's -4% less energy being radiated out with only a -1% drop in absolute temperature.

As NASA says, the radiant energy will always fall or rise faster than the temperature change. Meaning even though the increase in CO^2 will lead to a slight increase in the insulating value of the Earth's atmosphere towards outgoing radiation, the drop in radiated energy off the Earth's surface will always be greater, leading to a decrease in the Backscatter radiation cooling the earth further than just the drop in direct sunlight. A rise in the Backscatter radiation is impossible unless you can increase the temperature of the Earth.

THE EARTH IS WARMING, SO WHERE IS THIS WARMING COMING FROM?

Milankovitch's Eccentric Cycle

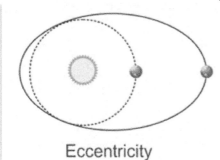

Eccentricity

Milutin Milankovitch was a Serbian genius Mathematician and

Astronomer born at the end of the 19 Century. It was he who first

notice that there were three distinct cycles throughout Earth's climatic

past. He was the first to understand that these periods were based upon the Earth's tilt, precession, and orbit. Important for our discussion here

is the last of these cycles, the orbital cycle Milankovitch named the

"Eccentric Cycle".

The Earth, as you no doubt have heard, will in 4.5 billion years

from now, drift off from the Sun never to return to it. The Earth's orbit

around the Sun is decaying. Each year the Earth gets a little farther from

the sun. However, it is not decaying gradually as you might think. The

Earth's orbit around the Sun has a wobble in it. Over a 100,000-year

cycle, which gets longer with the years, the Earth goes from having a

near-circular orbit around the Sun to an elongated one, then it goes back

to a near-circular orbit again. At the furthest point, the Earth is

calculated to be 2.78 million miles farther from the Sun. The Earth thus

has an Ice Age every 100,000 years when it is at its furthest distance

from the Sun. For 50,000 years the Earth wobbles away from the sun

and cools into an Ice Age, then the Earth wobbles back towards the Sun

and warms for the next 50,000 years.

Current scientific estimates state that the Earth has been moving

closer to the Sun for the past 45,000 years, and will continue to get closer to the sun, and so warm-up, for the next 5,000 years. As the Earth warms there is less and less cold areas to absorb the increasing heat and so the rate at which the Earth warms increases as the dwindling cold areas are overwhelmed by the increasing warmth of the Earth.

During the Little Ice Age, the Earth continued to get closer to the Sun but the Sun was in a Solar Minimum, reaching its trough during the Maunder Minimum (1645 - 1715). The Sun was putting out much less heat and light, so there was no increase in warmth for the entire 400 year-long Little Ice Age. When the Little Ice Age ended, the Sun returned to its usual output, but the Earth was now 400 years closer to the Sun, so the Earth got a big increase in solar radiation dumped on it all at once. At first the increased ice and snow and cold ocean waters left over from the Little Ice Age absorbed that added warmth, but over the next century that ice and snow melted, and the heat started to get the upper hand.

In 1971, the National Oceanographic and Atmospheric Administration announced that the surface water temperature in the North Atlantic rose for the first time since humans started to keep records. The North Atlantic had used up its extra cold leftover from the Little Ice Age and now would start to warm noticeably, leading to greater ice melt in the Arctic as North Atlantic waters flowed northward bringing its newly warmed surface waters into that region.

In the year 2012, the North Atlantic saw its most sever Hurricane season ever, ending with the biggest hurricane in more than 90 years, Hurricane Sandy.

Given the Coriolis effect in the northern hemisphere, with a record number of hurricanes in the North Atlantic, a record amount of warm North Atlantic water was pushed into the Artic, leading to record ice loss in the Artic Circle.

This led all the AGW so called "experts" to claim the end of the Arctic as we know it. Professing that this was going to happen every year. Of course, it didn't. The Ice pack in 2013 was 60% greater. But why?

The North Atlantic kicks out a Hurricane Sandy every 90 years or so. The earth's weather systems are energy systems. Energy cannot just come out of nowhere. The weather system has to slowly build up the energy for a really big hurricane season, and it just happens to take

90 years to do that. With the North Atlantic's energy gas tank emptied by the super hurricane season of 2012, 2013 saw the fewest hurricanes since humans started keeping records. Not only that, but the first hurricane of the season set the record for the latest first hurricane ever.

All scientists know these things, they know there is nothing that can be done to stop the Earth from warming over the next 5,000 years, they know about the 90-year hurricane cycle in the North Atlantic. They know we are going to go back to where the Earth was during the last interglacial age when the sea level was 22 to 24 feet higher than today, and Greenland was completely free of ice. This future cannot be altered, that the Liberals say that it can is a lie, and they know it. The Earth is warming, it has happened numerous times before and it will happen long after the human race is dead. Earth's orbital dynamics cannot be changed by eliminating Carbon dioxide emissions.

COSMO-CLIMATOLOGY AND CERN'S "CLOUD PROJECT"

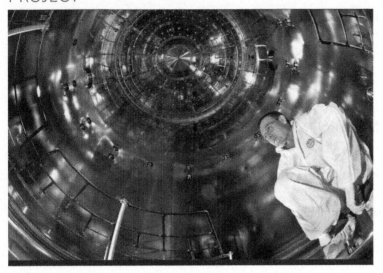

[CERN Project Director Jasper Kirkby inside the Cloud Chamber]

For those studying the link between solar activity and its potential effect on earth's climate, the interval of time since the year 1610, for which records of sunspots exist, is much too short to make any determinations of cause and effect certain. For earlier times the level of solar activity must be derived from other data sources. Fortunately for climate science, this information is stored in tree rings and ice cores in the form of "cosmogenic" isotopes.

Cosmogenic isotopes are radioactive nuclei resulting from collisions of energetic cosmic ray particles with the air molecules in our upper atmosphere. These collisions leave trace isotopes of Carbon 14 (C-14) in tree rings, and Beryllium-10 (Be-10) in ice cores. C-14 is a radioactive carbon molecule with a half-life of 5,730 years and is a well-used method to determine the age of wooden objects. The amount of C-14 produced depends strongly on the number of cosmic ray particles that reach the atmosphere. This number, in turn, varies with the level of solar activity. During times of high activity, the sun gives off more light and heat and produces more Sunspots. In addition, the Sun's magnetosphere also grows stronger and provides a more effective shield against these energetic cosmic particles, reducing cosmic ray interactions in the upper atmosphere.

When the intensity of the sun diminishes, the sun gets cooler and less bright, and it has fewer sunspots. In addition, the sun's magnetosphere weakens as well and so, allows more cosmic rays to enter the earth's atmosphere, creating more C-14 and Be-10 isotopes. These isotopes get locked away in tree rings and ice cores, leaving a roadmap of past cosmic-ray activity. Thus, we know higher solar activity leads to a stronger sun magnetosphere which blocks out more cosmic rays, leading to a lower production rate of C-14 and Be-10. While lower solar activity leads to a weaker sun magnetosphere and more cosmic rays entering into the atmosphere, leading to higher C-14

and Be-10 production rates.

During the Maunder Minimum (1645 to 1715), the coldest period in

human record, earth saw the highest production rate of C-14 and Be-10

in the earth's atmosphere. C-14 and Be-10 levels also matched

astronomic observation of sunspots during this time as historically low

numbers of sunspots were also recorded.   During the Maunder

Minimum, astronomers recorded the same number of sun spots over a

40-year period, as the sun normally has during the average solar year.

What we know from modern scientific equipment is that during

the 20th Century, the Sun's magnetosphere, which protects us from cosmic rays, **doubled in strength**. This resulted in dwindling numbers of C-14 and Be-10 isotopes in ice and tree rings laid down during this time confirming what our satellites and machines tell us. Fewer and fewer cosmic rays were penetrating the powerful solar magnetosphere.

So, what has cosmic rays not getting into the earth's atmosphere have to do with a warming earth?

Cosmic rays make clouds! That is why CERN's Cosmo-Climatology experiment is called the CLOUD PROJECT.

THE BEGINNING OF COSMO-CLIMATOLOGY.

Peer review articles published in the 50s, 60s, and 70s noted a rather precise link between C-14, and Be-10 in tree rings and ice cores that matched up well to historic climate variations. Large quantities of these isotopes were found when temperatures were low, and few found when temperatures were high. By the 70's it was being bandied about that these isotopes might be produced by cosmic radiation entering the atmosphere. It was proposed that the cosmic radiation may be creating "rarified plasmas" when it strikes particles in the upper atmosphere. These rarified plasmas it was thought, might be able to act as seedlings for clouds.

Enter Danish Physicist Henrik Svensmark, Director of the Center for Sun-Climate Research at the Danish National Space Center in Copenhagen Denmark. Svensmark was fascinated by this idea of cosmic radiation creating rarefied plasmas and did the basic math. Svensmark thought that this cosmic ray- rarified plasma-cloud seeding, might solve a mystery. How to reconcile the Sun's influence on climate, with the small 0.1% variations in the solar irradiance over a solar cycle measured by satellites. Svensmark realized that if true, this seeding of clouds by cosmic rays would create a solar-driven climate change mechanism much more powerful than mere changes in the solar irradiance. All he needed was money to build the experiment to prove that cosmic rays did produce the needed rarified seedlings.

Money for research, however would not be forthcoming. Svensmark approached CERN but CERN's Directors did not want to wade into the pool that was man-made global warming, and denied his request for funds.

The "consensus" was that mankind was making the earth warmer, and Svensmark's experiment if it proved to be true, would question that consensus, if not blow it out of the water. As such Svensmark found no backers. Svensmark returned to his lab and worked on his own to prove the essential role of creating these plasmas with a modest but affordable experiment.

The result of his work was announced at the 1996 COSPAR Space Science Meeting in Birmingham England. It was published the following year in the JOURNAL OF ATMOSPHERIC AND SOLAR-TERRESTRIAL PHYSICS, Volume 61, Issue 11, p. 795-797 (1997), under the title: VARIATION OF COSMIC-RAY FLUX AND

GLOBAL CLOUD COVERAGE-A MISSING LINK IN SOLAR-CLIMATE, by H. Svensmark and Eigil Friss-Christensen.

In 1998 Svensmark's work caught the attention of Jasper Kirkby, at CERN. Kirby approached CERN with an experiment to establish the cloud cosmic ray hypothesis, based upon the ionization Svensmark's experiment proved was happening, but again, CERN was still not interested in getting into the discussion of man-made climate change and denied funding.

After the second denial by CERN, Svensmark realized a recent discovery that low-level clouds are particularly sensitive to cosmic-ray variations suggested to Svensmark that a simple, and affordable experiment could be done at sea level. He called his experiment SKY. Svensmark continued this work in the basement of the Danish National Space Center and by 2005 Svensmark had found the causal mechanism by which cosmic rays can facilitate the production of clouds. The experiment showed that electrons released in the air by cosmic rays act as catalysts, significantly accelerating the formation of stable, ultra-small clusters of sulphuric acid and water molecules, which are the building blocks for cloud condensation nuclei. Svensmark created huge

numbers of microscopic droplets in the reaction chamber. The speed

and efficiency with which the electrons did their work of stitching

molecular clusters together took Svensmark completely by surprise.

Svensmark had discovered a climatologic mechanism that was

previously unknown in meteorology, which brought the cosmos and

cosmic rays into climate studies with a bang!

Svensmark's findings now forced CERN's hand. Politics were

politics, but CERN was a serious research institution dedicated to top-of-

the-line research. The Directors knew the danger they faced, the

possibility that nations might pull out of CERN and cut its funding, but

this was SCIENCE, not Politics. The SKY findings could not be denied,

and in 2006, CERN gave Kirkby clearance to start the **Co**smic **L**eaving

Outdoor **D**roplets Project (CLOUD Project).

In 2010, in a little lab at CERN, which no one talked about out of

fear of reprisal, the CLOUD Project was turned on and began collecting

data to no one's notice. Which suited the Directors at CERN just fine.

The less trouble CLOUD caused them the better.

Based at the Proton Synchrotron (PS) at CERN, CLOUD was the first time a high-energy physics accelerator was used to study atmospheric and climate science. During its first 5 years in service CERN's CLOUD Project had identified the vapors responsible for aerosol particle formation in the atmosphere. Cloud droplets cannot form without these aerosol particles, and so understanding what they are, how they are made, and how they work is vital to understanding cloud formation in the atmosphere. These vapors include Sulfuric Acid, Ammonia, Amines, and brand new to the discussion Biogenic Vapors from Trees (BVT)!

The BVTs were a big discovery and a total surprise. They had never even been discussed before as a possible method of seed production, yet experiments showed they play a major role in cloud formation.

Low-level clouds cover almost a quarter of the sky and exert a strong cooling effect on the surface. The 2% change in low cloud cover seen during the sun's 11-year solar cycle will vary the heat reaching the earth by an average of $1.2Wm^2$.

Given that the IPCC believes that added CO^2 has increase warming by 1.4Wm^2 since the 1850s, the 1.2Wm^2 variation due to decreased cosmic ray interaction with the atmosphere due to increases in the sun's magnetosphere, strongly suggests that cosmic rays may dominate climate change, not CO^2 variance.

It was long held that Sulfur Dioxide was necessary for aerosol particle formation, and because most sulfur dioxide in the atmosphere today comes from the burning of carbon fuels, computer climate models hold that there were fewer clouds in the pre-industrial world as there was no significant industrialization to produce the Sulfur Dioxide needed for aerosol particle formation. CLOUD has shown that that belief is incorrect. While most aerosols do require sulphuric acid, purely biogenic vapors from trees can also cause aerosol particle formation by themselves, without and Sulfur Dioxide being present. More dramatic than that was the CLOUD Project has shown that their formation rate is greatly enhanced by cosmic rays by a factor of 100!

When you look into the distance and see a blue haze, that haze is made of particles released by trees, rising up in the atmosphere. At altitude, these particles form into aerosol particles that can form clouds. The CLOUD Project shows that these biogenetic vapors can create clouds without any pollution being present. This is a totally new discovery that goes against past climate and cloud models. These Biogenetic Vapors from Trees have been shown by the CLOUD Project to dominate particle growth in unpolluted environments. More dramatic was the CLOUD Project showed that as these particles increase in size, they grow even more rapidly under cosmic ray bombardment, creating whiter, and fuller clouds. This makes these clouds super-efficient at absorbing and blocking out sun heat from reaching the earth's surface. Purely Biogenetic Nucleation is being shown as the primary factor in cloud formation in the Pre-industrial age, as all particles were being processed by cosmic ray ionization in the pure unpolluted air of earth's past. This means that there was a lot more clouds in earth's history than believed. This also means that all the past climate models are wrong. They have far too few clouds factored into their calculations and must be tossed out. Those models hold that Sulfur Dioxide Nucleation was the process by which particles were made, we know now that that is not correct, and so greatly underestimate cloud cover cooling in earth's past.

These results suggest that cosmic rays may have played a more important role in aerosol and cloud formation in pre-industrial times than in today's polluted atmosphere. This would lead to much greater swings in the amount of cloud coverage during the high and low webs in cosmic radiation entering the atmosphere due to increases and decreases in the sun's magnetosphere. With an increase in the magnetosphere, there would be a dip in the number of cosmic rays entering leading to a decrease in the number of clouds produced, and so, much more heat reaches the earth's surface. Likewise, as the sun's magnetosphere decreases in strength, there would have been more cosmic rays entering the atmosphere leading to an increase in cloud cover blocking out even more sun heat from reaching the earth making the earth even colder than the drop in solar output would make it.

With the Discovery of the role of the cosmos in earth climate more eyes have turned skyward. Cosmo-climatology has provided two new explanations for past earth climate that has puzzled scientists.

Researches have spotted a 32-million-year cyclic cooling pattern in the earth's history that matches up the two great cold periods in earth's climate history 2.3 billion and 700 million years ago. These cold periods match up with the earth's passage crosses the Mid-Plane of the Milky Way where cosmic rays are locally more intense. This would lead to many more clouds to block out solar heat greatly cooling the earth.

Also, Cosmo-climatology explains the great mystery of the young earth's heat. 4 billion years ago when the sun was young, it is estimated that it was only 25% as hot as today, yet sediments from that time show a wet world with liquid water. According to climate models based only upon the solar irradiance, the earth should have been a snowball covered in ice and snow. Revised solar models based upon modern data and Cosmo-climatology show a near absence of cosmic rays then. Thus, the earth had little to no clouds as the young sun's vigorous solar wind would have reduced the influx of cosmic rays to a small fraction of the present rate leaving the earth' without cloud cover to cool it. This could explain how the earth maintained a temperature able to accommodate liquid water in spite of the cool, young sun's lower temperature.

The biggest question however still remains unanswered. How much more cloud does an increase in cosmic radiation create? It will take more experiments before the CLOUD Project can answer that question with certainty. However, as the CLOUD Project accumulates more and more data, we get nearer to the definitive answer. Right now, we can only say cosmic radiation does play a pivotal a role, but how big has yet to be scientifically established.

What we can now say definitively is, it is NOT just how warm the sun is, it's about how much of that heat can get past the atmosphere and reach the earth that matters, and Cosmo-climatology can tell us how much heat is getting in.

Even the United Nations IPCC admit that clouds cool the earth and that more clouds cool the earth more, and fewer clouds cause the earth to get warmer. Thus, it is this balance, between the strength of the sun's rays and its magnetosphere that dominates earth temperatures. They work in reverse. More heat equals a stronger magnetosphere, and less cloud cover and an even warmer earth. Less heat equals a weaker magnetosphere, and more cloud cover and a cooler earth. When the sun is strong, not only does the earth get more sun, but it also sees fewer clouds giving it an extra boost of heat. While, when the sun gets weak and sends out less heat, the earth gets more clouds and so is made even colder.

THE FALLACY OF A RUNAWAY GREENHOUSE EFFECT, AND THE REAL TEMPERATURE OF VENUS. IT ISN'T AS HOT AS YOU THINK

The most common lie told by global warming advocates is that the human race is creating a Runaway Greenhouse Effect, just like Venus has.

Total Rubbish!

Remember, the atmosphere is like the insulation in your home. It prevents heat from moving across the insulation. The denser and thicker the insulation in your walls is, the less heat can cross the insulation. Four inches of insulation is okay, but twelve inches is better. Fluffy fiberglass insulation is okay, but thick foam insulation is better still.

Venus, unlike the Earth, has an atmosphere that is 95% CO_2, it is much denser. In addition, Venus' atmosphere is 50 kilometers thick. Over 5 times the thickness of the Earth's atmosphere. Venus' atmospheric insulation is five times thicker and denser! Which makes it a better insulator, and as Venus gets twice as much solar radiation than the earth, Venus needs that insulation.

Not only that, but Venus is completely covered in a thick, permanent blanket of CO_2 clouds that reflect so much of the Sun's radiation back out into space that Venus is the brightest planet in our solar system.

Russian VENERA landers which set down on Venus, tell us that Venus' surface temperature is 864F. Why is Venus so hot if so little heat passes through its thick cloud cover? Clouds on earth cool the earth during the day, so they should do the same thing on Venus, but still, Venus' temperature is 864F degrees (462C) while the Earth's average surface temperature is 57.2F degrees (14 C). Why the big difference in temperatures?

We know as NASA explains, a runaway Greenhouse effect is impossible as it violates the Stefan-Boltzmann Law. As any mass gets hotter, it will give off greater and greater amounts of heat, equal to the Fourth Power of its temperature checking any runaway effect. Why Venus is so hot is explained by two other laws at play on Venus, they are BOYLE'S LAW, and GAY-LUSSAC'S LAW? You might have learned Gay-Lussac's Law it in school as the PRESSURE/TEMPERATURE LAW.

In 1662, Robert Boyle published his work in which he noted that the

Pressure of a gas is inversely proportional to its volume.

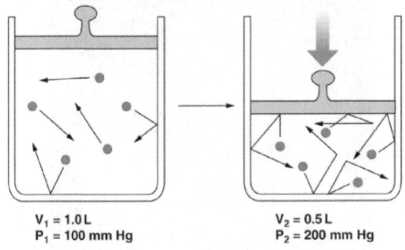

Boyle's Law: $P_1V_1 = P_2V_2$

Decreasing volume increases collisions and increases pressure.

$V_1 = 1.0\,L$
$P_1 = 100$ mm Hg

$V_2 = 0.5\,L$
$P_2 = 200$ mm Hg

Boyle noted that what we call Pressure, is merely the number of collisions between atoms in the container with the wall of the container. Atoms move in the container. It takes "X" amount of time for the atom to travel from one side of the container and collide with the other side and back again. But if you halve the sized of the container, then it takes only "1/2 X" for the atom to travel from one side to the other and back again. Thus, there are twice as many collisions when the volume is halved as when the volume is doubled.

Joseph Louis Gay-Lussac
1778 - 1850

Forty years later (1702), Joseph Louis Gay-Lussac discovered that the temperature of the gas was also directly related to its Pressure and wrote Gay-Lussac's Law.

Temperature is a measure of how many times the atoms of a gas collide with the thermometer used to measure its temperature. Just as the number of collisions with its container is the measure of the gases pressure as discovered by Boyle. Thus, both Temperature and Pressure are measures of the gases Kinetic Energy, (energy of motion). Any change in one creates an identical change in the other because they measure the same thing. One is measured with a manometer (pressure), and the other with a thermometer (temperature), but both are measuring

the same thing…the number of, and force of, the collisions the atoms of

the gas.

Like statute miles and kilometers both measure the same distance

but give different numbers. The same distance but different numbers are

due to the use of a different measuring system.

We know that you can change the Kinetic energy of a gas by either:

1. Compressing it into a smaller volume increasing the kinetic

    energy, or by enlarging the volume decreasing its kinetic energy;

    or

2. By burning a fuel and heating it, increasing its kinetic energy, or

    by refrigerating it and cooling it decreasing its kinetic energy.

Thus, we see, the temperature can be changed by external heating or

refrigeration, AND by reducing or increasing the volume of the gas.

Inflate your bicycle tires with a hand pump and the pump gets hot. The

pump gets hot because you've increased the atmospheric pressure of the ambient air with the pump by reducing its volume, thus you increased its kinetic energy and so heated the pump. This is heating by pressurization or Pressure Heat!

Pressure heat is also how a diesel engine works. It has no spark plug to ignite the fuel, instead, the air is compressed to 25 times the atmospheric pressure, which raises the air's temperature by a factor of 25 times the ambient air temperature. Pressure heat is also how Nature makes a sun.

Our solar system used to be filled with gas. No planets no sun. Over millions of years, the mass of the gas caused it to gravitationally coagulate at the center of our solar system. As a greater and greater weight of hydrogen gas was pressurized at the center of the solar system, the pressure got so high that it reached a temperature of Ten-million Degrees at the center of the gas and set off a fission reaction, and the sun was born.

Venus's atmosphere is 5 times thicker than the earth's and it is made out of heavier CO^2 gas. So common sense tells you that its atmospheric Pressure must be greater than on earth. So how much pressure is there on the surface of Venus? What did the Russian Venera landers measure when they set down on Venus surface?

90atms!

Ninety atmospheres of pressure versus one atmosphere of pressure on the earth's surface. This tells us that Venus' temperature must be 90 times that of the earth's temperature all other things being the same.

When we are speaking of earth heat at 1 atm, we're are speaking solely of Sun heat that passes through the atmosphere. At only 1atm, there is not enough pressure to make a significant difference in the kinetic energy of the surface heat. But Venus has 90atms of pressure at its surface plus the sun's heat that passes through the atmosphere. That is a lot of Pressure Heat on Venus's surface.

So, if we are to know the Solar Heat content of Venus' atmosphere, we need to **remove** all that Pressure Heat created by the 90atms. Thankfully, Gay-Lussac drew up the formula for doing just that, it is this:

P1/T1 = P2/T2.

This says that the Starting Pressure (P1) [in atmospheres] divided by the starting Temperature (T1) [in degrees Kelvin] is equal to the ending Pressure (P2), divided by the ending Temperature (T2).

If you prefer multiplication to fractions it is written this way:

P1 x T2 = P2 x T1

The starting Pressure times the ending Temperature is equal to the ending Pressure times the Starting Temperature.

Makes perfect sense, right?

Sure, it does, so, let's do the math and find out what Venus's real temperature is minus all that Pressure Heat.

We want to know the temperature of Venus at 1 atm, so its ending pressure (P2), will be 1atm like the Earth. Its' starting pressure (P1), is 90atms, and its starting temperature (T1) is 864F (462.2C, or 735.4 degrees Kelvin (K)).

You can go here and use a calculator to do the math yourself:

https://www.omnicalculator.com/physics/gay-lussacs-law

The default on the calculator uses "Pascals" (Pa) and Kelvin for the Initial Pressure and Temperature. Just hit the down arrow to the right of the "Pa" and "K" and select "standard atmospheres (atm)" and Fahrenheit or Celsius if you would like to use those measuring temperature scales. Don't worry, the calculator will convert F and C into K for you.

(Note if you do the math yourself, you MUST use degrees Kelvin.)

Now, enter 90 into the "Initial Pressure" (Venus' atmospheric Pressure), and 1 into the "Final Pressure" (earth's atmospheric pressure). Then enter 864F into the "Starting Temperature". In the formula we need to isolate "T2" by moving P1 to the other side of the formula which reverses the function from multiplication to division and we get this: T2 = (P2T1)/P1.

When we put in the numbers it looks like this:

T2 = (1atm x 735K)/90atm

1 x 735.4 = 735.4.

735.4 divided by 90 is 8.17K.

The website calculator gives us a temperature of **-445F**. This number has been rounded up from the exact number of **-444.9622** which your calculator will give you if you do the math yourself.

So, T2 = 8.171K which converts into **-264.979C**, or **-444.9622F**!

Thus, we see that Venus real temperature, its temperature minus all the Pressure Heat that is created by a heavy CO^2 atmosphere that is 5 times thicker than earth's atmosphere is a bitter cold **-444.9622F**. That's not hot at all is it? All that CO^2 in Venus's atmosphere, blocks out so much of the Sun's heat that it is ice cold on Venus, even though, at the top of Venus atmosphere, it gets twice as much heat as the earth does.

This conversion shows you the importance of having an education in Physics. The liar Al Gore and his lying acolytes pushing Man-Made CO^2 warming, have been lying about Venus' temperature and a runaway greenhouse effect for decades. But you didn't know they were lying because you did not have a Physics education.

So, we know the atmosphere of Venus blocks out almost all the heat of the sun, but the tremendous amount and weight of the atmosphere of Venus creates 90atms of atmosphere pressure, and so increases its kinetic energy by a factor of 90 times.

Are you wondering what the temperature of the earth would be at 90atms? Let's use the calculator to find out.

The earths average temperature is 57.2 degrees F (14 C). So, the starting Pressure is 1 atm, the starting temperature is 57.2F, and the Final Pressure is 90 atms.

At 90atms the calculator tells us that earth's surface temperature is **46,059F**!!! GOOD GOD THAT'S HOT!!!

Far hotter than Venus at only 864F.

And remember Venus gets **twice as much sun heat** from the sun as the earth does, so its temperature at 90atms, should be twice the earth's temperature at 90atms, or **92,118F** degrees, yet it is a mere 864F, 1/106th of what it should be at 90atms. Venus temperature is nowhere close to what it should be given how much solar heat it gets at the top of its atmosphere. All that CO^2 cloud cover blocks out the Sun's heat so effectively Venus is ice cold when you remove all the pressure heat.

116

ARE HUMAN ACTIVITIES WARMING THE EARTH?

Yes, we are warming the earth.

Humans have cut down most of the forests and trees in North American and Europe, and are now doing it in South America and have already done it in most of Africa. Trees absorb the sun's heat and keep the earth cooler. European civilization has replaced these forests with asphalt-covered streets and cement buildings creating a "Heat Island Effect" in our cities.

Trees absorb sun heat keeping the earth's surface cool, reducing its albedo (its reflectance). Also, trees draw moisture from under the soil and evaporate it off their leaves, adding to the moisture level in the air, increasing rainfall which cools the atmosphere at its lower altitudes, and at high altitudes, as water vapor, it helps to block out sun heat and keep the earth cooler. In addition, the lighter, dryer air rises to higher altitudes, which causes wind patterns to change.

Man's terraforming is having an impact on the Earth's climate. Not to mention the animals and insects we are displacing, or have outright destroyed by destroying their habitat.

In the United States, we have destroyed 75% of the Countries wetlands. Think of that, **75%**! You should be shocked at the stupidity of this destruction! That is a lot of wetlands no longer evaporating water into the air. Not to mention that water is no longer there for the animals and insects to drink. That cannot be good, and it isn't.

The bee population is disappearing because we have destroyed so much of the Bee's habitat. Bees pollinate 80% of our edible plants. When they are dead and gone, we will be hard-pressed to feed ourselves.

People also burn gasoline and other fossil fuels putting heat into the atmosphere helping to warm it.

So, we are **destroying the natural cooling systems** nature has developed and were **adding heat** to the atmosphere as well!

Human civilization is definitely warming the earth there is no doubt about that. But the CO^2 and Sulfur Dioxide we are adding to the atmosphere is counteracting some of our warming influences, by blocking out more Sun heat from passing through the atmosphere and warming the earth. CO^2 remember, does NOT warm the planet. Like with clouds on a sunny day, it keeps out much more heat than it keeps in and so CO^2 has a **Net Cooling effect**, not a warming one. Thus, there is no such thing as a Greenhouse Gas. No ThRM warms more than it cools

118

the earth. All the so-called "Greenhouse Gases" have a Net Cooling

effect on the Earth's temperature, they don't warm it.

A FINAL WORD ABOUT THE IPCC

The United Nations International Panel on Climate Change (IPCC), is NOT a Scientific body. It has no Laboratories, and it does no scientific research nor testing. It does not even attempt to scientifically establish anything it claims in its reports which are largely laughable attempts at scientific reporting. This failure to meet scientific regimen is, to a large degree, because it is NOT a body of scientists. Few of its members have any scientific background at all. They are largely all laypeople just like the average American or Hollywood star or starlet. These Laypersons, read articles in scientific fields they have no education in, and then make judgments about what they have read deciding on which is true and valuable. They are no different than the Stars and Starlets of Hollywood, they are idiots. It is impossible to understand a scientific field without an education in it. Laypeople simply cannot understand what they are reading. The idea that such laypeople think that they can evaluate a scientific report is laughable.

In the introduction to this book, I made clear my attitude toward those who make claims about any field of study without an education in it. These members of the IPCC are just more Idiots who think you don't need an education in a scientific field to be able to understand that field. They are more ego than intellect.

Here is what 45 of the Scientists who have worked on the IPCC have to say about the institution and the IPCC's reports:

Dr. Robert Balling: *"The IPCC notes that "No significant acceleration in the rate of sea level rise during the 20th century has been detected." This did not appear in the IPCC Summary for Policymakers."*

Dr. Robert C Balling, Jr. is a professor of geography at Arizona State University, and the former director of its Office of Climatology. His research interests include climatology, global climate change, and geographic information systems. Balling has declared himself one of the scientists who oppose the consensus on global warming, arguing in a 2009 book that anthropogenic global warming "is indeed real, but relatively modest", and maintaining that there is a publication bias in the scientific literature.

Dr. Lucka Bogataj: *"Rising levels of airborne carbon dioxide don't cause global temperatures to rise.... temperature changed first and some 700 years later a change in aerial content of carbon dioxide followed."*

Dr. Lucka Bogataj, is a joint recipient of the Nobel Peace Prize in 2007, is one of Slovenia's pioneers in researching the impact of climate

121

change, and she regularly informs the general public of her findings.

She is a full professor and teaches at the Biotechnical Faculty, while also

lecturing at the Faculty of Mathematics and Physics and at the Faculty of

Architecture

Dr. John Christy: *"Little known to the public is the fact that **most of**
the scientists involved with the IPCC do not agree that global warming
is occurring. Its findings have been consistently misrepresented and/or
politicized with each succeeding report."*

Dr. John Christy is a climate scientist at the University of Alabama in

Huntsville (UAH) whose chief interests are satellite remote sensing of

global climate and global climate change. He is best known, jointly with

Roy Spencer, for the first successful development of a satellite

temperature record. He is the Distinguished Professor of Atmospheric

Science and Director of the Earth System Science Center at the

University of Alabama in Huntsville. He was appointed Alabama's state

climatologist in 2000. For his development of a global temperature data

set from satellites. He was awarded NASA's Medal for Exceptional

Scientific Achievement, and the American Meteorological Society's

"Special Award." In 2002, Christy was elected Fellow of the American Meteorological Society.

Dr. Rosa Compagnucci: *"Humans have only contributed a few tenths of a degree to warming on Earth. Solar activity is a key driver of climate."*

Dr. Rosa Compagnucci is retired but she continues advancing in her past line of research. Four years ago he worked at the Department of Atmospheric and Ocean Sciences, Universidad de Buenos Aires and was Principal Research in the Argentina Research Council CONICET. Rosa does research in Climatology, Meteorology, and Paleoclimatology. Their most recent publication is 'RELATIONSHIP AMONG A SUPERNOVA, A TRANSITION OF POLARITY OF THE GEOMAGNETIC FIELD AND THE PLIO-PLEISTOCENE BOUNDARY'.

Dr. Richard Courtney: *"The empirical evidence strongly indicates that the anthropogenic global warming hypothesis is wrong."*

Dr. Richard Courtney is a Technical Editor for CoalTrans International (journal of the international coal trading industry) who lives in Epsom,

Surrey (UK). In the early 1990s, Courtney was a Senior Material Scientist of the National Coal Board (also known as British Coal) and a Science and Technology spokesman of the British Association of Colliery Management. Member of the European Science and Environment Forum. Acting as a technical advisor to several U.K. MPs and mostly-U.K. MEPs

Dr. Judith Curry: *"I'm not going to just spout off and endorse the IPCC because I don't have confidence in the process."*

Dr. Judith Curry is an American climatologist and former chair of the School of Earth and Atmospheric Sciences at the Georgia Institute of Technology. Her research interests include hurricanes, remote sensing, atmospheric modeling, polar climates, air-sea interactions, climate models, and the use of unmanned aerial vehicles for atmospheric research. She is a member of the National Research Council's Climate Research Committee. After publishing over a hundred scientific papers and co-editing several major works, Curry retired from academia in 2017.

Dr. Robert Davis*: "Global temperatures have not been changing as state-of-the-art climate models predicted they would. Not a single mention of satellite temperature observations appears in the IPCC Summary for Policymakers."*

Dr. Robert Davis is a Professor of Climatology at the University of Virginia's Department of Environmental Sciences. Davis received his Ph.D. in 1988 from the University of Delaware. His research contributions include the development of a system for measuring the power of Nor'easters. In his studies of global warming, he has suggested that it may manifest more by milder winters than by hotter summers, and predicted that its effects on the human population will not be severe.

Dr. Willem de Lange: *"In 1996 the IPCC listed me as one of approximately 3000 "scientists" who agreed that there was a discernible human influence on climate. I didn't. There is no evidence to support the hypothesis that runaway catastrophic climate change is due to human activities."*

Dr. Willem de Lange's Position: Senior Lecturer Echo-sounding, Earth Sciences, Faculty of Science and Engineering, University of Waikato.

Field: Earth and ocean sciences, focus on coastal oceanography. An earth scientist and lecturer at the University of Waikato, was born in the Netherlands and moved with his family to New Zealand when he was 18 months old. Since then, he has stayed put in Hamilton. He did his Bachelor of Science, master's, and Ph.D. at the University of Waikato and is now a Senior Lecturer in the Earth and Ocean Sciences Department there.

Dr. Chris de Freitas: *"Government decision-makers should have heard by now that the basis for the long-standing claim that carbon dioxide is a major driver of global climate is being questioned; along with it the hitherto assumed need for costly measures to restrict carbon dioxide emissions. If they have not heard, it is because of the din of global warming hysteria that relies on the logical fallacy of 'argument from ignorance' and predictions of computer models."*

Dr. Chris de Freitas is a climate scientist. He was an associate professor in the School of Environment at the University of Auckland. De Freitas, born in Trinidad, received both his Bachelor's and his Master's at the University of Toronto, Canada, after which he earned his Ph.D. as a

Commonwealth Scholar from the University of Queensland, Australia. During his time at the University of Auckland, he served as deputy dean of science, head of science and technology, and for four years as pro-vice-chancellor. He also served as vice-president of the Meteorological Society of New Zealand and was a founding member of the Australia–New Zealand Climate Forum.

Dr. Oliver Frauenfeld: *"Much more progress is necessary regarding our current understanding of climate and our abilities to model it."*

Dr. Oliver Frauenfeld research activities include a broad range of topics in climate variability and climate change. I focus primarily on surface-atmosphere interactions, over both the land and the oceans. One of these research areas investigates changes in Arctic and high-altitude environments; specifically, the interactions between frozen ground (permafrost and seasonally frozen areas) and other cryospheric variables in the high latitudes of Eurasia, with the overlying atmosphere.

Dr. Peter Dietze: *"Using a flawed eddy diffusion model, the IPCC has grossly underestimated the future oceanic carbon dioxide uptake."*

127

Dr. Peter Dietze Independent energy advisor and scientific climate and carbon modeler; official reviewer, Intergovernmental Panel on Climate Change; Bavaria, Germany. Independent energy advisor and scientific climate and carbon modeler; official reviewer, Intergovernmental Panel on Climate Change; Bavaria, Germany.

Dr. John Everett: *"It is time for a reality check. The oceans and coastal zones have been far warmer and colder than is projected in the present scenarios of climate change. I have reviewed the IPCC and more recent scientific literature and believe that there is not a problem with increased acidification, even up to the unlikely levels in the most-used IPCC scenarios."*

Dr. John Everett is a marine biologist who has worked with NOAA and the Intergovernmental Panel on Climate Change (IPCC), and manages the UN Atlas of the Oceans; he is currently president of the consulting firm Ocean Associates, Inc." He was a Member of the Board of Directors of the NOAA Climate Change Program from its inception until he left NOAA. He led several impact analyses for the Intergovernmental Panel on Climate Change (IPCC) from 1988 to 2000, while a NOAA employee. The reports were reviewed by hundreds of government and academic scientists as part of the IPCC process."

Dr. Eigil Friis-Christensen: *"The IPCC refused to consider the sun's effect on the Earth's climate as a topic worthy of investigation. The IPCC conceived its task only as investigating potential human causes of climate change."*

Dr. Eigil Friis-Christensen received a Magisterkonferens (Ph.D. equivalent) in Geophysics from the University of Copenhagen in 1971. In 1972, he was a geophysicist at the Danish Meteorological Institute. His interest in solar activity began in August, in his tent, when he experienced an extreme solar storm.

Dr. Lee Gerhard: *"I never fully accepted or denied the anthropogenic global warming concept until the furor started after NASA's James Hansen's wild claims in the late 1980s. I went to the [scientific] literature to study the basis of the claim, starting with first principles. My studies then led me to believe that the claims were false."*

Dr. Lee Gerhard is a retired geologist from the University of Kansas. His profile at Thomasson Partner Associates, Inc. describes him as an

Honorary Member of the American Association of Petroleum

Geologists, past president and Honorary Member of that society's

Division of Environmental Geosciences, an Honorary Member of the

Association of American State Geologists, and an Honorary Member of

the Kansas Geological Society.

Dr. Indur Goklany: *"Climate change is unlikely to be the world's most*

important environmental problem of the 21st century. There is no signal

in the mortality data to indicate increases in the overall frequencies or

severities of extreme weather events, despite large increases in the

population at risk."

Dr. Indur Goklany is a science and technology policy analyst for the

United States Department of the Interior, where he holds the position of

Assistant Director of Programs, Science and Technology Policy. He has

represented the United States at the Intergovernmental Panel on Climate

Change (IPCC) and during the negotiations that led to the United

Nations Framework Convention on Climate Change. He was a

rapporteur for the Resource Use and Management Subgroup of Working

Group III of the IPCC First Assessment Report in 1990, and is the author of Clearing the Air (1999), The Precautionary Principle (2001), and The Improving State of the World (2007).

Dr. Vincent Gray: *"The [IPCC] climate change statement is an orchestrated litany of lies."*

Dr. Vincent Gray was a New Zealand chemist, and a founder of the New Zealand Climate Science Coalition. Gray was awarded a Ph.D. in physical chemistry by the University of Cambridge. He commented on every publication of the Intergovernmental Panel on Climate Change, with 1,898 comments on the 2007 Report.

Dr. Mike Hulme: *"Claims such as '2500 of the world's leading scientists have reached a consensus that human activities are having a significant influence on the climate' are disingenuous ... The actual number of scientists who backed that claim was only a few dozen."*

Dr. Mike Hulme Professor of Human Geography in the Department of Geography at the University of Cambridge. He was formerly a professor of Climate and Culture at King's College London (2013-2017) and of Climate Change in the School of Environmental Sciences at the University of East Anglia (UEA). Hulme served on the United Nations'

Intergovernmental Panel on Climate Change (IPCC) from 1995 to 2001. He also contributed to the reports of the IPCC.

Dr. Kiminori Itoh: *"There are many factors which cause climate change. Considering only greenhouse gases is nonsense and harmful." "Warming fears are the "worst scientific scandal in the history...When people come to know what the truth is, they will feel deceived by science and scientists."*

Dr. Kiminori Itoh is a Japanese award-winning environmental physical chemist who contributed to the U.N. IPCC AR4 climate report. He received his Ph.D. in industrial chemistry from the University of Tokyo in 1978. He has written four books on this issue. He also has a patent on a method of sunspot number anticipation, and contributed to the IPCC AR4 as an expert reviewer."

Dr. Yuri Izrael: *"There is no proven link between human activity and global warming. I think the panic over global warming is totally unjustified. There is no serious threat to the climate."*

Dr. Yuri Izrael was a vice-chairman of the Intergovernmental Panel on Climate Change (IPCC) until September 2008, when the new bureau was elected. Izrael was the former chairman of the Committee for Hydrometeorology. He also served as director of the Institute of Global Climate and Ecology, which is a part of the Russian Academy of Sciences. He was the first vice-president of the World Meteorological Organization and helped develop the World Weather Watch.

Dr. Steven Japar: *"Temperature measurements show that the climate model-predicted mid-troposphere hot zone is non-existent. This is more than sufficient to invalidate global climate models and projections made with them."*

Dr. Steven Japar, a Ph.D. atmospheric chemist who, was part of the Intergovernmental Panel on Climate Change's (IPCC) Second (1995) and Third (2001) Assessment Reports, and has authored 83 peer-reviewed publications and in the areas of climate change, atmospheric chemistry, air pollutions, and vehicle emissions.

Dr. Georg Kaser: *"This number [of receding glaciers reported by the IPCC] is not just a little bit wrong, it is far out by any order of magnitude ... It is so wrong that it is not even worth discussing."*

Dr. Georg Kaser is a South Tyrolean glaciologist Dr. Georg Kaserand is considered one of the most influential climate researchers worldwide. He worked twice as lead author on the Intergovernmental Panel on Climate Change (IPCC), the United Nations' World Council of Nations.

Dr. Aynsley Kellow: *"I'm not holding my breath for criticism to be taken on board, which underscores a fault in the whole peer review process for the IPCC: there is no chance of a chapter [of the IPCC report] ever being rejected for publication, no matter how flawed it might be."*

Dr. Aynsley Kellow is a climate skeptic at the School of Government, University of Tasmania. Aynsley Kellow was an IPCC reviewer to Working Group II of AR4.

Dr. Madhav Khandekar: *"I have carefully analyzed adverse impacts of climate change as projected by the IPCC and have discounted these claims as exaggerated and lacking any supporting evidence."*

Dr. Madhav Khandekar is a former research scientist from Environment Canada and is presently on the editorial board of the Journal of Natural Hazards (Kluwer). He is an environmental consultant on extreme weather events and a scientist with the Natural Resources Stewardship Project. He has worked in the fields of weather and climate for nearly 50 years and has published more than 120 papers, reports, and book reviews and, a monograph on ocean surface wave analysis and modeling (Springer-Verlag 1989). Khandekar is one of the external reviewers for the Intergovernmental Panel on Climate Change's 1997 Fourth Assessment Report.

Dr. Hans Labohm: *"The alarmist passages in the IPCC Summary for Policymakers have been skewed through an elaborate and sophisticated process of spin-doctoring."*

Dr. Hans Labohm studied economics and economic history at the Municipal University of Amsterdam. After his military service, from 1967 he worked for the Ministry of Defense at the Dutch Permanent Representation to NATO in Brussels. In 1971 he joined the Foreign Service and was sent to Sweden. After returning to the Netherlands in 1974, he worked in various positions at the Ministry of Foreign Affairs: from 1978 as Deputy Policy Planning Advisor. From 1987 to 1992 he was Deputy Permanent Representative of the Netherlands to the

Organization for Economic Cooperation and Development (OECD) in Paris. Since September 1992, he has been affiliated with the Clingendael Institute as a guest researcher and advisor to the Board of Directors. He has regularly published in Het Financieele Dagblad, NRC Handelsblad, de Volkskrant, the Internationale Spectator, and Liberal Reveil, among others. From 2002 he worked as a columnist on the American website of 'Tech Central Station'. After years of blogging for The Daily Standard (DDS), he has been writing for Climategate.nl and Yalta since June 2015.

Dr. Andrew Lacis*: "There is no scientific merit to be found in the Executive Summary. The presentation sounds like something put together by Greenpeace activists and their legal department."*

Dr. Andrew Lacis NASA-GISS Lacis_AGU_2011 National Aeronautics and Space Administration (NASA AST, CLIMATE & RADIATION STUDIES). Andrew A. Lacis received his B.A. in Physics in 1963, M.S. in Astronomy in 1964, and Ph.D. in Physics in 1970, all from the University of Iowa. He was selected for the NASA traineeship program, a program established by NASA to encourage graduate students in the pursuit of scientific research and study. While a graduate student, he also did research in astrophysics and astronomy in Japan at the University of Kyoto, and at the University of Tokyo. Following his Ph.D., he was Instructor in Astronomy at the University of Iowa. In 1972, he teamed up with James E. Hansen for post-doctoral research in planetary atmospheres at the Goddard Institute for Space Studies (GISS).

Dr. Chris Landsea: *"I cannot in good faith continue to contribute to a process that I view as both being motivated by pre-conceived agendas and being scientifically unsound."*

Dr. Chris Landsea is an American meteorologist, formerly a research meteorologist with the Hurricane Research Division of the Atlantic Oceanographic and Meteorological Laboratory at NOAA, and now the Science and Operations Officer at the National Hurricane Center. He is a member of the American Geophysical Union and the American Meteorological Society.

Dr. Richard Lindzen: *"The IPCC process is driven by politics rather than science. It uses summaries to misrepresent what scientists say and exploits public ignorance."*

Dr. Richard Lindzen is an American atmospheric physicist known for his work in the dynamics of the middle atmosphere, atmospheric tides, and ozone photochemistry. He has published more than 200 scientific papers and books. From 1983 until his retirement in 2013, he was Alfred P. Sloan Professor of Meteorology at the Massachusetts Institute of Technology. He was a lead author of Chapter 7, "Physical Climate Processes and Feedbacks," of the Intergovernmental Panel on Climate Change's Third Assessment Report on climate change. He has criticized the scientific consensus about climate change and what he has called *"climate alarmism."*

Dr. Harry Lins: *"Surface temperature changes over the past century have been episodic and modest and there has been no net global warming for over a decade now. The case for alarm regarding climate change is grossly overstated."*

Dr. Harry Lins is a Scientist Emeritus (Hydrology) with the U.S. Geological Survey. During his years at USGS, his work spanned several Earth science disciplines, including coastal processes, surface water hydrology, and hydroclimatology. Although most of his career was spent conducting research, he managed the USGS Global Change Hydrology Program from 1989 to 1997, and served as Co-Chair of the Intergovernmental Panel on Climate Change (IPCC) Hydrology and Water Resources Working Group for the IPCC First Assessment Report. In 1999, he and USGS colleague David Wolock developed "WaterWatch", the Nation's first website depicting maps and graphs of water resources conditions in near real-time. Lins currently serves as President of the World Meteorological Organization (WMO) Commission for Hydrology.

Dr. Philip Lloyd: *"I am doing a detailed assessment of the IPCC reports and the Summaries for Policy Makers, identifying the way in which the Summaries have distorted the science. I have found examples of a summary saying precisely the opposite of what the scientists said."* *"The quantity of CO2 we produce is insignificant in terms of the natural circulation between air, water and soil. I am doing a detailed* *assessment of the UN IPCC reports and the Summaries for Policy Makers, identifying the way in which the Summaries have distorted the science. Google natural CO2 vs man-made CO2 for the real facts."*

Dr. Philip Lloyd UN IPCC co-coordinating lead author, [Nuclear Physicist] and Chemical Engineer, and author of more than 150 refereed publications. Philip Lloyd's Professional details- Honorary Research Fellow: Energy Research Centre: University of Cape Town, Cape Town; Fellow: SA Academy of Engineering; Chair: VAF: Chemical & Allied Industries Association; Fellow: SA Chemical Institute (SACI)

Dr. Martin Manning*: "Some government delegates influencing the IPCC Summary for Policymakers misrepresent or contradict the lead authors."*

Dr. Martin Manning was the Founding Director of the New Zealand Climate Change Research Institute at Victoria University of Wellington, established to build better interactions between science, policy, and society on climate change issues. From 2002 to 2007, Martin was Director of the Intergovernmental Panel on Climate Change (IPCC) Working Group I Technical Support Unit that produced the Fourth

Assessment Report on climate change for governments. He has produced over 50 papers in peer-reviewed scientific literature and been an author and review editor for several of the major IPCC reports. Martin has worked in several countries but spent most of his life in New Zealand where he led research on greenhouse gases, atmospheric chemistry, and other aspects of climate change science over the last thirty years. In 2008, Martin became an Officer of the New Zealand Order of Merit for his services to climate change science.

Dr. Patrick Michaels: *"The rates of warming, on multiple time scales, have now invalidated the suite of IPCC climate models. No, the science is not settled."*

Dr. Patrick Michaels is a past president of the American Association of State Climatologists and was program chair for the Committee on Applied Climatology of the American Meteorological Society. He was a research professor of Environmental Sciences at the University of Virginia for 30 years. Michaels was a contributing author and is a reviewer of the United Nations Intergovernmental Panel on Climate Change, which was awarded the Nobel Peace Prize in 2007.

Dr. Nils-Axel Morner: *"If you go around the globe, you find no sea level rise anywhere."*

Dr. Nils-Axel Morner is the former head of the paleogeophysics and geodynamics department at Stockholm University. He retired in 2005. He was president of the International Union for Quaternary Research (INQUA) Commission on Neotectonics (1981–1989). He headed the INTAS (International Association for the promotion of cooperation with scientists from the New Independent States of the former Soviet Union) Project on Geomagnetism and Climate (1997–2003). He is a critic of the IPCC and the notion that the global sea level is rising.

Dr. Johannes Oerlemans: *"The IPCC has become too political. Many scientists have not been able to resist the siren call of fame, research funding and meetings in exotic places that awaits them if they are willing to compromise scientific principles and integrity in support of the man-made global-warming doctrine."*

Dr. Johannes Oerlemans is a Dutch climatologist specialized in glaciology and sea level. He is a professor of meteorology in the Faculty of Physics and Astronomy at Utrecht University.

Dr. Roger Pielke: *"All of my comments were ignored without even a rebuttal. At that point, I concluded that the IPCC Reports were actually intended to be advocacy documents designed to produce particular policy actions, but not a true and honest assessment of the understanding of the climate system."*

Dr. Roger Pielke is an American political scientist and professor, and was the director of the Sports Governance Center within the Department of Athletics at the Center for Science and Technology Policy Research at the University of Colorado Boulder. He previously served in the Environmental Studies Program and was a Fellow of the Cooperative Institute for Research in Environmental Sciences (CIRES) where he served as Director of the Center for Science and Technology Policy Research at the University of Colorado Boulder from 2001 to 2007. Pielke was a visiting scholar at Oxford University's Saïd Business School in the 2007-2008 academic year.

Dr. Paul Reiter: *"As far as the science being 'settled,' I think that is an obscenity. The fact is the science is being distorted by people who are not scientists."*

Dr. Paul Reiter is a professor of medical entomology at the Pasteur Institute in Paris, France. He is a member of the World Health Organization Expert Advisory Committee on Vector Biology and Control. He was an employee of the Center for Disease Control (Dengue Branch) for 22 years. He is a specialist in the natural history, epidemiology, and control of mosquito-borne diseases such as dengue fever, West Nile fever, and malaria. He is a Fellow of the Royal Entomological Society. Reiter says he was a contributor to the third IPCC Working Group II (Impacts, adaptation, and vulnerability) report, but resigned because he "found [himself] at loggerheads with persons who insisted on making authoritative pronouncements, although they had little or no knowledge of [his] specialty". After ceasing to contribute he says he struggled to get his name removed from the Third report.

Dr. Murry Salby: *"I have an involuntary gag reflex whenever someone says the science is settled. Anyone who thinks the science is settled on this topic is in fantasia."*

Dr. Murry Salby is an American atmospheric scientist who focused on upper atmospheric wave propagation for most of his early career, and who more recently argued against aspects of the scientific consensus that human activity contributes to climate change. He has written two textbooks, Fundamentals of Atmospheric Physics (1996), and Physics of the Atmosphere and Climate (2011). The latter textbook, building on his first book, offers an overview of the processes controlling the atmosphere of Earth, weather, energetics, and climate physics.

Dr. Tom Segalstad: *"The IPCC global warming model is not supported by the scientific data."*

Dr. Segalstad conducted research, publishing, and teaching in geochemistry, mineralogy, petrology, volcanology, structural geology, ore geology, and geophysics at the University of Oslo and at Pennsylvania State University. His current research interests include general geochemistry (the chemistry of the Earth), metallogenesis (how mineral deposits and ore deposits form), igneous petrogenesis (how magmatic rocks form), and carbon dioxide and the "greenhouse effect" (how carbon dioxide cannot cause "global warming"). He is past head of the Natural History Museums and Botanical Garden of the University of Oslo and currently a member of several international and national

145

professional working groups and committees, including an expert reviewer for the Intergovernmental Panel on Climate Change.

Dr. Fred Singer: *"Isn't it remarkable that the Policymakers Summary of the IPCC report avoids mentioning the satellite data altogether, or even the existence of satellites — probably because the data show a slight cooling over the last 18 years, in direct contradiction of the calculations from climate models?"*

Dr. Fred Singer is an atmospheric and space physicist founded the Science and Environmental Policy Project (SEPP) and the Nongovernmental International Panel on Climate Change (NIPCC). He served as professor of environmental sciences at the University of Virginia, Charlottesville, VA (1971–94); distinguished research professor at the Institute for Space Science and Technology, Gainesville, FL (1989–94); chief scientist, U.S. Department of Transportation (1987–89); vice-chairman of the National Advisory Committee for Oceans and Atmosphere (NACOA) (1981–86); deputy assistant administrator for policy, U.S. Environmental Protection Agency (1970–71); deputy assistant secretary for water quality and research, U.S. Department of the

Interior (1967– 70); founding dean of the School of Environmental and Planetary Sciences, University of Miami (1964–67); first director of the National Weather Satellite Service (1962–64); and director of the Center for Atmospheric and Space Physics, University of Maryland (1953–62).

Dr. Hajo Smit: *"There is clear cut solar-climate coupling and a very strong natural variability of climate on all historical time scales. Currently I hardly believe anymore that there is any relevant relationship between human CO2 emissions and climate change."*

Dr. Hajo Smit a former member of the UN IPCC committee who reversed his belief in man-made warming to become a skeptic.

Dr. Richard Tol: *"The IPCC attracted more people with political rather than academic motives. In AR4, green activists held key positions in the IPCC and they succeeded in excluding or neutralizing opposite voices."*

Dr. Richard Tol is a professor of economics at the University of Sussex. He is also a professor of the economics of climate change at the Vrije Universiteit Amsterdam. He is a member of the Academia Europaea. Tol was a coordinating lead author for the IPCC Fifth Assessment Report

Working Group II: Impacts, Adaptation, and Vulnerability. Tol said in March 2014 that he had withdrawn from the writing team for the Summary for Policy Makers of the report in September 2013, citing disagreement with the profile of the report which he considered too alarmist and putting too little emphasis on opportunities to adapt to climate changes.

Dr. Tom Tripp: *"There is so much of a natural variability in weather it makes it difficult to come to a scientifically valid conclusion that global warming is man-made."*

Dr. Tom Tripp is Professor of Management, Rom Markin Endowed Leadership Chair in Business, and Senior Associate Dean for Academic Affairs at the Carson College of Business at Washington State University. He previously taught at the Kellogg School of Management at Northwestern University and at the Sauder School of Business at the University of British Columbia.

Dr. Gerd-Rainer Weber: *"Most of the extremist views about climate change have little or no scientific basis."*

Dr. Gerd-Rainer Weber Ger-Rainer undertook undergraduate and graduate studies in atmospheric sciences at the Free University of Berlin, during which time he was a Fulbright and Indiana University Scholar. Further study in America gained him an M.Sc. degree in atmospheric sciences from the University of Michigan. He returned to the Free University of Berlin to study for his Meteorology Ph.D. in conjunction with the Max-Planck Institute of Aeronomy.

Dr. David Wojick: *"The public is not well served by this constant drumbeat of alarms fed by computer models manipulated by advocates."*

Dr. David Wojick a Ph.D. in the philosophy of science and mathematical logic from the University of Pittsburgh and a B.S. in civil engineering from Carnegie Tech. He has been on the faculty of Carnegie Mellon University and the staffs of the U.S. Office of Naval Research and the Naval Research Lab.

Dr. Miklos Zagoni: *"I am positively convinced that the anthropogenic global warming theory is wrong."*

Dr. Miklos Zagoni is a physicist and science historian at Eotvos Lorand University, Budapest, now a governmental adviser. He is a well-known science writer in Hungary. He participated in the Hungarian Academy of Science's climate change project and was the expert-reporter of three documentary films on that project. His list of publications, interviews, papers, and book chapters on the issue is more than 200 items.

Dr. Eduardo Zorita: *"Editors, reviewers, and authors of alternative studies, analysis, interpretations, even based on the same data we have at our disposal, have been bullied and subtly blackmailed."*

Dr. Eduardo Zorita is a Spanish paleo-climatologist. As of 2010, he is a Senior Scientist at the Institute for Coastal Research, GKSS Research Centre in Geesthacht, Germany, where he has worked since 1996. Zorita was a contributing author to the Fourth Assessment Report of the IPCC, and is review editor of the journal Climate Research.

REFERENCES

Almeida, J. Scholbesberger, S. et al, 2013, Molecular Understanding of Sulphuric Acid-Amine Particle Nucleation in the Atmosphere. *Nature,* 502, 359-363.

Bard, E., and Raisbeck, G.M., Yiou, F., Jouzel, J., 1997, *Earth Planet Sci let. 150, 453.*

Arrhenius, S., 1896 *Philo. Mag. & J. of Science,* Series 5, Volume 41, April 1896, pages 237-276

Bard, E. and Raisbeck, G.M., Yiou, F., Jouzel, J., 2000, *Tellus B*, 52, 985.

Bazilevskaya, G.A., 2000, *Space Sci. Rev.,* 94, 25.

Beer, J., 2000, *Space Sci. Rev.* 94, 53.

Beer, J. and Blinow, A., Bonani, G. Hofmann, H.J., Finkkel, R.C., 1990 *Nature* 347, 164.

Bertrand, C., and Loutre, M.F., Crucifix, M., Berger, A., 2000 *Tellus A*, 54, 221

Bond, G et al, 2001 *Science* 294 2130-6.

Chalmers, M. 2016 *Science* (Oct.) CLOUD Experiments Sharpens Climate Predictions.

CLOUD, 2000 CERN/SPSC 2000-021 SOSC-P317.

Cubasch, U., and Voss, R., 2000, *Space Sci, Rev.,* 94, 185.

Eddy, J.A., 1997 *Science* 192 1198-1202.

Elrick, M. and Hinnov, L.A., 2006 *Palaeogeog., Palaeoclim., Palaeoecol.*

Ebisuzaki, W. J., 1997 *Clim.*10,2147.

Dio:10.1016/j.palaeo. 2006.08.008.

Fligge, M., Solanki, S.K., 2000, *Geoph. Res. Let.,* 27, 2157.

Friis-Christensen, E. and Lassen, K., 1991 *Science* 254 698-700.

Frohlich, C., 2000, *Space Sci. Rev.,* 94, 15.

Gillies, J. 2016 *Nature* (May) CLOUD Experiment Shows Pre-Industrial Skies were Cloudier than We Thought.

Herschel, W., 1801 *Phil. Trans. Tay Soc.* 91 256-318.

Hills, S., 2013 *Nature* (Oct.) Cows, Clouds, and Climate.

Houghton, J.T. et al, (ed) 2001 Climate Change 2001: The Scientific Basis (Cambridge U. Press).

Hoyt, D.V., Schatten,K.H., 1992 *J. Geoph.Res.* 179, 189.

Kirkby, J. 2013 *Nature* (Oct.) CLOUD Shines new light on Climate Change.

Kirkby, J. 2014 *Science* (May) CLOUD Sheds new Light on Cloud Formation.

Kirkyb, J. 2016 *CERN* (May) CLOUD Points to a Cloudier Pre-Industrial Climate.

Kirkby, J. 2020 *Nature* (May) CLOUD Discovers New Mechanism that

Drives Winter Smog Episodes in Cities.

Knie, K. et al., 2004 *Phys. Rev. Lett.* 93 171103.

Kristjansson, J.E. and Kristiansen, J. 2000 *JGR* 105 11851-63.

Krivova, N.A., Solanki, S.K. in A. Wilson (ed.): 2003, *Solar Variability as an Input to the Earth's Environment,* ESA SP-535, European Space Agency, 275.

Larkin, A., Haigh, J.D., and Djavidnia, S., 2000 *Space Sci. Rev.,* 94, 199.

Lean, J., and Beer, J., Bradley, R., 1995, *Geophys. Res. Let.,* 26, 759.

Lockwood, M. et al., 1999 *Nature* 939 437-439.

Lopez, N. 2019 *CERN* (Oct.), From Cosmic Rays to Clouds.

Lu, Q. B., 2013, Cosmic Ray Driven Reaction and Greenhouse Effect of Halogenated Molecules: Culprits for Atmospheric Ozone Depletion and Global Climate Change. *Inter. J. of Modern Phys. B., Vol. 27, No. 17*

Mann, M.E., and Bradley, R.S., Huges, M.K., 1999, Northern Hemisphere Temperatures During the Past Millenium: Inferences, Uncertainties, and Limitations, *Geophys. Res. Let.,* 26, 759.

Mann, M.E., Jones, P.D., 2003, *Geophys Res. Let.,* 30, DOI 10.1029/2003GL017814.

Marcos, R. del la F, and Marcos, C de la F, 2004 *New Astronomy* 10 53-66.

Marsh, N. and Svensmark, H., 2000 *Phus. Rev. Let.* 85 5004-5007.

Marsh, N., Svensmark, H., 2000, *Space Sci. Rev.,* 94, 215.

Mitchel, J.H.B., et al., 2001 Chapter 12 pp695-735 in Houghton et al. 2001.

Raymo, M.E., 2006 *Science* 313 492-495.

Reid, G.C., 2003, *Clim. Change,* 37, 391.

Riccobono, F. and Schobesberger, S., Scott, C.E., Dommen, J., Ortega, I.K., et al., 2014, Oxidation Products of Biogenic Emissions Contribute to Nucleation of Atmospheric Particles, *Science, Vol.344, Issue. 6185, pp. 717-721. DOI: 10.1126/science.1243527.*

Rocha-Pinto, H.J. et al., 2000 *A7A* 358 869-885

Rottman, G., 1999, *J.Atm. Terr. Phys.,* 61, 37.

Royer, D. et al., 2004 *GSA Today* 14 (March) 4-10.

Shackleton, N.J., 2000 *Science* 289 1897-1902.

Shaviv, N., 2002 *Phys. Rev. Let.* 89 051102.

Shaviv, N., 2003a *New Astronomy* 8 39-77.

Shaviv, N., 2003b *JGR* 108 (A12) 1437 dio:10.1029 /2003JA009997.

Shaviv, N. and Veizer, J. 2003 *GSA Today* 13 (July) 4-10.
Solanki, S.K., Fligge, M., 1998, *Geoph. Res. Let.,* 25, 341.
Solanki, S.K., Fligge, M., 1999, *Geophys. Res. Lett.* 26 2465.
Solanki, S.K., and Schussler, M., Fligge, M., 2000, *Nature*, 408, 445.
Solanki, S.K., and Schussler, M., Fligge, M., 2002, *Astron. Astrophys.,* 383, 706.
Soreghan, G.S. et al., 2005 *GeoSystems: Probing Earth's Deep-Tie Climate & Linked Systems,* NSF workshop report downloadable at http:..www.geosystems.org.
Stott, P.A., and Jones, G.S., Mitchell, J.F.B., 2003, *J. of Climate* 16, 4079.
Svensmark, H. 1998 *Phys. Rev. Let.* 81 5027-5030.
Svensmark, H. 2003 http://arxiv.org/abs/physics/0311087
Svensmark, H. 2006a *Astron. Nachrichten* 327 866-870.
Svensmark, H. 2006b *Astron. Nachrichten* 327 871-875.

Svensmark, H. 2007 eprint (in preparation).
Svensmark, H. and Calder, N. 2007 *The Chilling Starts; A New Theory of Climate Change* (Icon Books, Thriplow, Cambridge).
Svensmark, H. and Friis-Christensen, E. 1997 *j. Atmos. & Solar-Terrest. Phys.* 59 1225-1232.
Svensmark, H. et al., 2007 *Proc. Ray. Soc. A* (doi:10.1098/rspa.2006.1773 online Nov. 2006).
Usoskin, I.G., and Solanki, S., Schussler, M., Mursula, K., Alanko, K., 2003, *Phys. Rev. Lett.,* 91(21), 211101, dio:10.1103/*PhysRevLett*.91.211101.
Usoskin, I.G. and Schussler, M., Solanki, S.K., Mursula, K. Alanko, K., 2004, *Astron. Astrophys.,* 413, 745.
Usoskin, I.G. and Schussler, M., Solanki, S.K., Mursula, K., *Proceedings of the 13th Cambridge Workshop on Cool Stars, Stellar Systems and the Sun* European Space Agency 5-July-2004 "Solar Activity over the last 1150 years: does it correlate with Climate?"
Van Loon, H., Labitzke, K. 2000, *Space Sci. Rev.,* 94, 259.
Wagner, G. et al., 2001 *J. Geophys. Res.* 106 3381-3388.

Made in the USA
Monee, IL
17 March 2021